Falconry in Arabia

Mark Allen

Foreword by Wilfred Thesiger

Illustrated by
Mary-Clare Critchley-Salmonson

Orbis Publishing · London

FRONTISPIECE

Peregrine on manqalah

ACKNOWLEDGMENTS

Black and white photographs by Wilfred Thesiger
Colour photographs by Peter Whitehead
Maps by Creative Cartography Limited

First published in Great Britain by
Orbis Publishing Limited, London 1980

Printed Letterpress and Litho in Great Britain by
Hazell Watson & Viney Limited
Aylesbury, Bucks

ISBN 0 85613 013 3

Contents

To David Bourke and Roger Upton
who gave me a love and knowledge
of hawks and falconry.

Preface

I COULD NOT HAVE COME TO WRITE THIS BOOK WITHOUT THE HELP OF MANY friends both at home and in the many states of today's Arabia. As a token of my gratitude, not only to them but to their families and wider circles of friends, I want to mention Hasan Muri' al-Sheikh, Sheikh Naif Hadithah al-Khuraishah, the late Sheikh Lawrence Trad al-Sha'alan, Sheikh Surur Muhammad Al Nahayyan, Sa'id Salman Abu 'Adhirah, Ahmad Juma' al-Zarif and Burgess 'Abd al-Hamid Abu Turki.

I have never carried a camera in the desert and so must thank Wilfred Thesiger and Peter Whitehead for the use of their fine photographs. I have been lucky also in the help of an old friend, Mary-Clare Critchley-Salmonson. In her paintings and drawings for this book she has captured the personalities of trained hawks in the fine tradition of English falconry artists.

I am grateful to Professor Robert Sergeant and the editors of *Arabian Studies* for their kind permission to draw in the glossary on material which I first used with Dr Rex Smith in an article in that journal. Readers familiar with the work of François Viré and Dietrich Moller will know how much I owe to them in my summary of medieval hawking in the appendix.

Nonetheless, in spite of these debts, the opinions in this book are my own and if there are faults they are mine only. I hope that a friend from his self-imposed anonymity will approve the result. He did so much to encourage me.

Finally, my wife always supported the idea of this book – as I thank those to whom it is dedicated, I must also thank her.

Cairo 1979

Geographical map of the Middle East

Foreword

Falcons have always had a fascination for me. As a boy I went frequently to the Elan valley in central Wales, and there I spent hours watching peregrines coming on and off their eyries, enthralled by the rapid beats of their wings and their wild chattering cries. Of birds, only the sight and sound of wild geese have for me a comparative fascination.

I had always hoped one day to go hawking. I was fortunate, for in 1949 and 1950 I accompanied Sheikh Zayid of Abu Dhabi when he went forth to hawk in the desert. For a month or more we rode for long hours on superb camels, slept on the ground in the open, fed on the hares and bustards we had taken – half-a-dozen in one day if we were lucky – an exacting and rewarding experience in confronting an immemorial past. Today the old ways are gone. No sheikh any longer rides a camel. Instead they go hawking in cars and even use shot guns. They cover vast distances and by these methods have virtually exterminated the Macqueen's bustard in Arabia, as they have already exterminated the oryx, and now, in luxuriously appointed jet aircraft, they go even further afield to Iraq, Iran and Pakistan.

Mark Allen, an Arabic scholar and a dedicated falconer, has written a fascinating and absorbing book on Arab falconry. His yearning for the vanished past is everywhere apparent, but his account bridges the gap between yesterday and today. He first describes the Bedu in their tribal desert setting and then relates them to the Arab society of the Gulf that has taken their place, the product of sudden unimaginable wealth. In their passion for falconry he sees one of the few interests that still links the values of today with the past in which they grew up.

This book really is a mine of information about Arab methods of falconry – if only I could have read it before I went hawking with Sheikh Zayid, how greatly would I have profited.

Wilfred Thesiger

بِسْمِ اللَّهِ الرَّحْمَنِ الرَّحِيمِ

"يَسْأَلُونَكَ مَاذَا أُحِلَّ لَهُمْ قُلْ أُحِلَّ لَكُمُ
الطَّيِّبَاتُ وَمَا عَلَّمْتُمْ مِنَ الْجَوَارِحِ مُكَلِّبِينَ
تُعَلِّمُونَهُنَّ مِمَّا عَلَّمَكُمُ اللَّهُ فَكُلُوا مِمَّا أَمْسَكْنَ
عَلَيْكُمْ وَاذْكُرُوا اسْمَ اللَّهِ عَلَيْهِ
وَاتَّقُوا اللَّهَ إِنَّ اللَّهَ سَرِيعُ الْحِسَابِ"

صدق الله العظيم

(آية ٤ سورة المائدة)

They will ask you what is made lawful for them. Say: those things which are sound are lawful for you, and the prey of those creatures of the chase which you have trained like dogs, teaching them as God has taught you. Eat of what they catch for you and make mention of God's name over it, and fear God: God is swift to reckon.

The Koran.

OPPOSITE PAGE *Haggard peregrine*

Introduction

THE QUEST FOR OIL HAS UNLOCKED THE OLD FORTRESS OF ARABIA. IN many parts its people are now outnumbered by foreigners, newcomers whose money has made cities of remote settlements which fifty years ago were known only to explorers, a handful of political agents and servicemen. Industries and services stand up against the emptiness like dragon's teeth sown by the early pioneers. Strangeness and change have penetrated everywhere and the Arabs, in self-conscious defence against the tide, have had little option but to look for an anchorage in their own traditions as a way to preserve some continuity – some authentic personality – with which to face their new world.

In this search for an identity their love of hawking has played a part, and today falcons appear on cap badges and currency, as the emblems of countless companies and firms, even as regular accessories in advertisements to convey, apparently with every success, the atmosphere and feeling of Arabia. For certainly with the camel, the Arab horse, the black hair tent and the saluki, the hawk is a symbol of the desert Arab's way of life and it remains a feature of it which he has cherished through all the traumatic changes of the last few generations. In going hawking the Arab returns to the desert, to the forge which made him what he is.

Hawking is a form of hunting – using birds of prey to take wild quarry. The falconer's vocation is to manage the hawk in such a way that she (hawks in general are referred to in the feminine) will come to accept him as a spectator at her hunting. He must diet her and care for her. He must find her quarry and stalk it so that she has a fair chance when slipped to fly it. Her role when trained is to consummate the carefully laid plans of the falconer; and in his judgement in setting about these lies the art. The element of sport is introduced by the independent plans of hawk, weather and quarry.

Since the beginning of time, hawks and falcons have won man's attention. Their independence, imposing appearance and their mastery of flight have caught his imagination and he has painted, moulded and carved their likenesses, mummified them and written their praise. The alliance of falcon and falconer was a natural culmination. She is a specialist and thus amenable to exploitation: the falconer only wants her to practise her natural skills and she, like many other specialists, is little concerned with anything else.

Where hawking began and who was the first falconer were the favourite subjects of speculation in the introductions to medieval Arabic treatises on falconry. But these questions are made impossible to answer by the range of early evidence of falconry occurring from China to Mesopotamia. Though they knew of the sport, the Greeks and Romans did not practise it themselves, and records of falconers in Europe only became more frequent with the contraction of Roman influence. By the Middle Ages hawking was in full swing as the refined and highly technical pursuit of kings, emperors and caliphs from the Atlantic to Japan.

In Europe falconry remained pre-eminent as a sport till the seventeenth century. It was the only method of killing birds in flight and an honourable way of catching those which were edible. But then it was eclipsed by the development of guns and its popularity finally collapsed when these could be used to shoot birds on the wing. In England, the enclosure of land broke up much of the open ground once ideal for hawking and the preservation of game for shooting put hawks on the list of vermin. Their numbers suffered drastically in consequence. More enlightened attitudes after the Second World War finally led to a reprieve from this persecution. But there was another threat to come – the increased use of agricultural insecticides. The hawk as the last link in the food chain was found to take up accumulated doses of these poisons, causing the production of thin-shelled and infertile eggs. Strict legislation to remedy these ugly effects was passed before the damage was too great. There has been a recovery and it is interesting to see that many of the energies once devoted to sport now feed the passion for conservation. Hawks are back in vogue and their doyen, the peregrine, has become a figure of some charisma. Falconers, however, who have always thought this, are now often viewed with suspicion; their interests are regarded as a possible tax on the wild population. But they just manage to hold their own in conservationists' circles by asserting their ancient lore and knowledge of hawks. Their special position and value to the understanding of these birds is still recognized for the time being.

In the Middle East, things have followed a different pattern but the present situation is no less critical. There, hawking has a long and wide history but it survives today principally in the form in which it was practised by the bedouin, the nomadic tribesmen of the desert. Certainly there were other falconers – in medieval Baghdad and Damascus the sport had a grand following – but their tradition is now dead in Arabia and recorded only in manuscripts and miniatures: lost with a vanished world of pomegranates and sherbet. That branch of hawking is really a separate subject but there is a brief account of it in the appendix. It bears little relation, save language, to the sport of the desert which is what survives and is the subject of this book.

In the past the seclusion of Arabia ensured the survival of hawking in the desert. Developments which adversely affected the sport in other parts of the world had no access to this remote part – it continued very much as it had begun. In Arabia, even more than was the case in medieval Europe, which was so much richer in food supplies, game taken by the Arab's trained hawk was an important addition to his thin diet. Wild hawks were trapped by rich and poor alike and were flown through the winter months when quarry became available. The Arabs flew their hawks because they wanted to hunt and eat. They did not go hunting because they wanted to see their hawks fly. This is an important distinction to bear in mind as it underlies the techniques and apparent ruthlessness of desert hawking.

But that is not to say that it was a mechanical or dour business. In any part of the world there is more to hunting than pest control or pot filling, and for the bedouin the hawk's victory over its quarry was a feat of courage and strength in which they felt able to share through their alliance with the conqueror. In the hawk's graceful restraint at rest and her grim hardness in the field the bedouin saw qualities which were for him among the criteria of honour in a tribal society. Her dark eyes, flashing wings and the hold of her blooded hands were the stuff of poetry when he returned to the camp from a hunting expedition.

In the last twenty years enormous increases in the numbers of cars and firearms in the desert have tipped what was previously a natural balance over to the advantage of the hunter. Ever-declining quarry has become more and more the prerogative of the powerful and, now that their social status has been underpinned by the riches of oil, the sums they pay for hawks have soared. Nowadays it is a dedicated falconer who will prefer to train for himself a hawk he has trapped rather than realize its cash potential by offering it to a sheikh. So as a popular sport hawking has declined, but the position of its patrons still gives it a social significance. The trapping of hawks, their management or flying, remain important occupations for many tribesmen. Hawking is a source of income and it gives them contact with their sheikhs or rulers in a day when the old easy relations are otherwise increasingly rare.

Hawking has become mechanized and to an extent commercialized. It is the self-consciously Arab pastime of people whose culture has been flooded with the limelight of foreign ways and interests. The Arabs, however, do not shun this glare. With all the surface self-confidence of all remote peoples, they suffer none of the inhibitions about field sports which now affect us in the West. A demonstration of this was the decision of Sheikh Zaid, the President of the United Arab Emirates, to hold an international conference of falconers in 1976. The idea of the project caused no small alarm among the mainly town Arab and expatriate officials who were charged with the preparations for it. No such event had been held on this scale before.

I was living in Abu Dhabi at that time and was approached for ideas on a scheme that a world history of falconry be brought out in time for the conference which was scheduled for that December. Absolutely unable to convince the officials of the immensity of such a task, I took refuge in a copy of Harting's *Bibliotheca Accipitraria* and suggested that at least it could act as a reading list for their researches. Printed in 1891 it catalogues three hundred and seventy-eight works in nineteen languages ranging from Swedish and Provençal to Turkish and Japanese. Reasonably, as people of the sown (as opposed to the desert), the officials and organizers had no idea that what to them was simply the bedouin's pastime was but a branch of a great and broad-limbed tree. Later at the conference it emerged that little was known among the foreign experts and academics of what the desert Arabs' hawking actually amounted to. Save for articles and brief accounts in travellers' books, it has been unrecorded in the West. Now, if the ominous forecasts of the conservationists are not mistaken, it will not be long before hawking in Arabia is a thing of the past, following to extinction the wildlife and quarry which are its life blood. This is the branch of hawking about which I want to write.

Very much a personal record, this can scarcely amount to a treatise, still less a definitive work. To me, a falconer and then an Arabist, it is a souvenir: an attempt to set out what I have been taught by Arab falconers and what I have yet to learn before it is too late.

At home in England I began keeping and training hawks when I was fourteen, but my first encounter with hawking in the East was years later in

Map showing the principal tribes which fly hawks

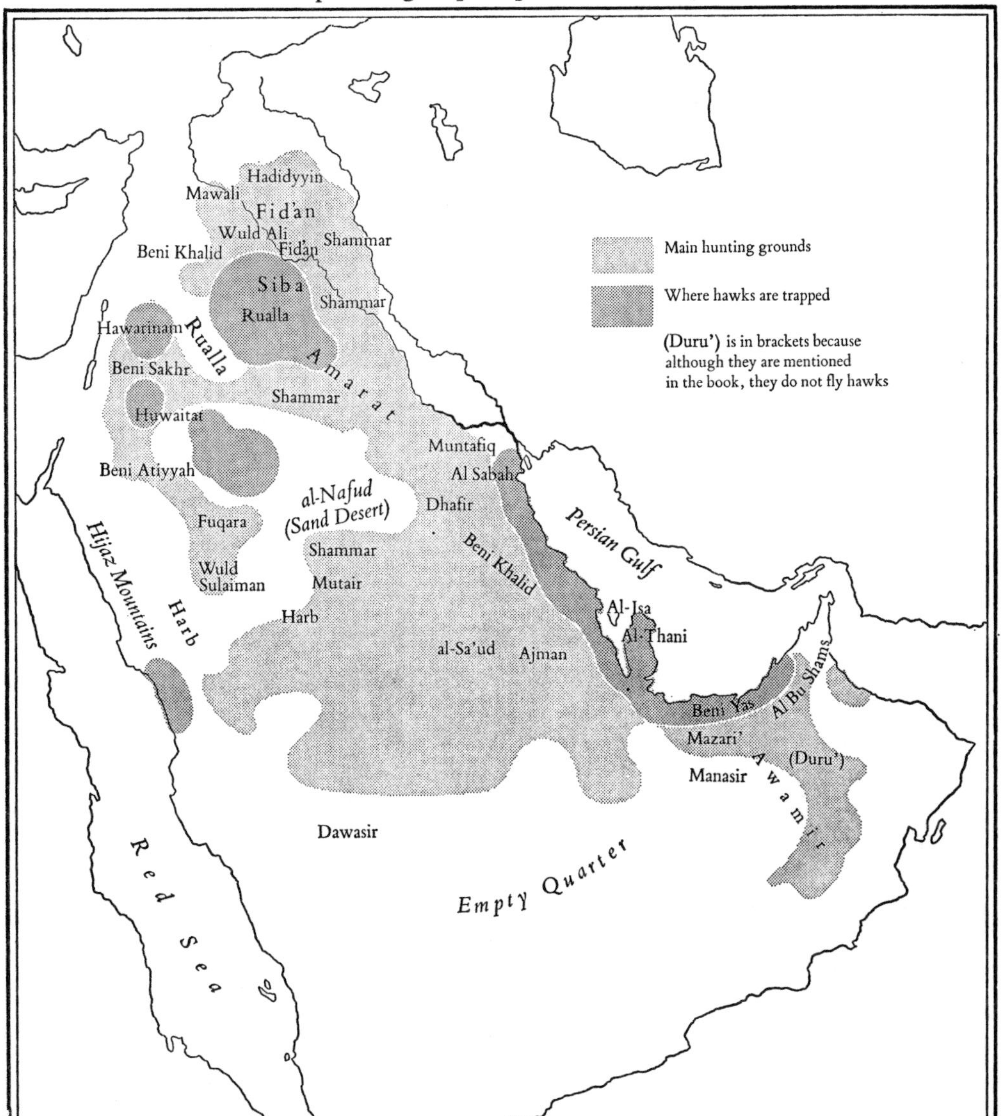

Turkey. A chance reference in the journal *Ibis* led me to the Black Sea coast. Here falconers still fly sparrowhawks at the quail which arrive on migration from southern Russia in their thousands. From a decision to change my course at Oxford from Classics to Turkish, it was a short step to Arabic though it was a year before I set foot in an Arab country. A short vacation course in the Lebanon was my first opportunity. But in the bazaars of Beirut falconers and hawks were only a distant rumour and I was directed to Damascus.

Damascus – the traditional starting point for the great caravan of pilgrims on their annual way to Mecca clustered together for protection against the bedouin, the terminus of Lawrence of Arabia's Revolt in the Desert and the city of caliphs and St Paul; after the arid mountain pass of the ante Lebanon the greenery of the gardens which surround the city was dazzling. But this soon gave way to the dust and disappointments of a town struggling on a chassis for which spares had not been made for ten or fifteen years. Kinglake, journeying the other way a hundred and twenty odd years before, had been exhilarated at his crossing back into the Lebanon and the West by the release of leaving behind him the stagnation of the East and rejoining the world of 'wheels, steam buzzing – and a slashing pace'. Damascus at first sight seemed to have made one supreme effort at modernization, bought in all the ugliness of machines and concrete, and then relapsed, sunk back into a torpor of making do.

After walking a while and some false starts in gun shops and taxidermists, the charm of a city so old and a people so versed in pleasing made their inevitable effect: endless small glasses of black tea, the deep shade of the bazaar shot through with pencils of dusty light and above, above the haystacks of baskets and the tapping out of copper and brass trays the size of tractor wheels, the siren call to prayer, shimmering across the hot roofs and the stone walls to ricochet off another minaret and be lost in another call. In a more practical corner of the market, away from the cherries and the silks, I found a squint-eyed man who had linnets, buntings and a thrush for sale, with hens and ducks besides. He told me that I should work back again up the street of parsley, coriander and dates, to a place by a mosque where there was a bus and the bus would take me to a village where I would find a hawk. One man on the bus spoke a little more English than my few words of Arabic. He was the schoolmaster and thus treated with some respect – a protection for me.

The fifty or so kilometres ran like an age under rusted splintering mountains and out onto a plain punctuated along the course of the road by a ruined khan or a whitewashed village. At the schoolmaster's own village we were dropped at a camouflage net of bougainvillaea, jasmine and vines. It should have concealed a huge gun but in fact shaded his house. After a large lunch, which his family's modesty and sense of hospitality determined I should eat alone, he called in a circle of friends. A turquoise and chrome mid-fifties Buick apparently stuffed with laundry disgorged five Arabs of hard and business-like features. Their only hostage to vanity in their severe white robes were heavy gold watches and cuff links, rings and cigarette lighters. They joined our cross-legged circle on the floor and watched me over the rims of the coffee cups, a ritual of welcome. They were trappers who caught up the hawks on their autumnal migration and then sold them to sheikhs in Saudi Arabia and the Gulf (hence the gold) and handled every year as many hawks as most European falconers would hope to see in a lifetime.

With cigarettes and tea they warmed and after hours of talk rushed us off in the Buick and an equally elderly Chevrolet to have supper at another house. There we sat, in a stone courtyard with an orange tree, late into the night under the stars, talking of snares, hoods and jesses, what manner of birds were the houbara and the Scottish grouse, and all the arcana of falconry. Heavy with a supper of curds, rice and mutton, we walked home to the schoolmaster's house under a cold moon set in a navy-blue sky. The village was silent except for dogs barking in the distance where the desert began and swept away unbroken till the Euphrates and Baghdad. It was my twenty-first birthday and I had made contact with men who shared what had always been with me a passion.

The surest thing I had understood was that I must return and a year later I did, to meet up again and to follow new leads further into the desert. By this time my aim was to get as far as I could from the English-speaking circles of Amman where I was living. I wanted to immerse myself in the language of the desert. Elegant people can propel themselves into foreign society with introductions from home. I had to pick my way through the perplexities of bedouin politeness with only a knowledge of hawks.

The Syrian trappers of the year before, though acknowledging tribal origins, were settled in stone houses. They cultivated gardens and spoke of the desert in terms of some respect. In Jordan, however, I was meeting bedouin who, though their sheikhs had large houses and their tribes centred on the villages which had grown up around these, still thought of the open wastes as home. Nearly all were armed, and they sat in great rings in the tents, watching with immobile faces framed in the red and white check of their head cloths. The intricacies and standards of their courtesy were daunting. At meals only the right hand could be put into the huge tray of rice and mutton which would be carried in by two negro servants, the descendants of slaves; the sole of the foot had to remain concealed in one's cross-legged sitting; direct questions about a person's past or even the whereabouts of one absent were taboo – a habit presumably derived from the need to protect each other from the ears of a vengeance-seeker in a blood feud. The thick black-hair tent held the heat of the fire to an extraordinary degree, and in spite of their forbidding aspect the tribesmen caught one up in the free openness of their welcome. The sheikhs were sometimes a little more cagey.

I remember arriving at the home of one sheikh and telling him that I had heard that he might be able to help me get a falcon. 'Is that so?' he replied. 'Well, if you could tell me how to catch one, it would follow that I might.' His amusement at my attempts to stutter the gutturals of Arabic and the procedures for trapping a hawk was such that I had to repeat them as a party piece for every newcomer who appeared at the tent. Another, who had such a fierce reputation among the town dwellers of Amman that I was begged not to take the risk of visiting his encampment, received my absurd cabaret act in stony silence. But late at night, when I was being put into a car for the three-and-a-half hour return to Amman, he pushed in with me the presents of a peregrine falcon and a young male saker. He told me to return when I could.

From such beginnings grew friendships and adventures which over the years led to more and always to an ever-increasing sense of gratitude. I was thankful that these beginnings and whatever I later learned of the Arab should have been built on an interest we shared and not one borrowed by either side. An Arab falconer living in the Liwa oasis, on the edge of the Empty Quarter, once said to me: 'In the past we were dependent on, yet masters of, our camels and their lore. Yet we were happy. What a dilemma that our happiness in days to come may lie in learning to become mechanics.'

Strange things always invite judgments, but I hope the reader will remember that the practices which I describe are those of Arabia and not of Europe or America, unless noted to the contrary. Our priorities differ and, while some of these practices were once common in the West, they have for the most part been superseded by other methods. In many countries, many of them have become illegal as society's perceptions of cruelty and fairness have changed.

OPPOSITE PAGE *Haggard lanner*

Arabs and Arabia

'These Arabs Mahommet was born among are certainly a notable people.'
Carlyle.

There are inevitably problems of scene setting and definition in trying to describe only a part of a strange way of life. What foreigners most usually notice in Arabia is the urge to modernize, expressed in the Western idiom of cement. Many live there for long periods without any contact at all with the society of the desert. It is a land full of contradictions.

Just to sketch the context in which the Arab goes hawking is to enter a minefield of reservations: his world is changing so fast that even most recent memories seem out of date. In giving a picture only of the present, the perspective of tradition is lost and perhaps its influence understated. But to write simply of the past would be to draw a scene unrecognizable to today's traveller in Arabia and conceal the thread of continuity which does in fact remain.

The barriers are formidable and not the least of these is the difficulty of language. Hawking in the West is thought of as a sport; but in Arabic there are no equivalent words for that apparently simple notion. There is a word for sport in the sense of gymnastics or exercise, and another meaning a hobby or pastime cultivated as an amateur. But with neither of these would it be possible to explain the idea of what is for us an essential element in the excitement of hunting: a sporting chance. So when we speak of falconry as the sport of the desert Arab, we are already making some assumptions which he does not share.

This dilemma of how to give a general but fair impression arises the very moment one first thinks of Arabs and Arabia. Is a Lebanese just as much an Arab as a native of Mecca? And where, in relation to the United Arab Emirates or 'the Palestinian homeland', is Arabia?

I shall use 'Arabs', as the nomads do, as the simplest word for the people of the desert, and mean no others. 'Take off his hood, let him watch the Arabs and be content!' was the straightforward advice given me by a falconer during a trip to Oman, when I was having some difficulty with a restless peregrine in the circle round the fire. As a bedouin, he had used 'the Arabs' quite unself-consciously to mean 'people' – there were no other foreigners apart from myself. This use was an old habit and in the past had given the word the special meaning of 'nomad' or 'bedouin', until the Arabic-speaking peoples of the Ottoman Empire began to search for and claim a uniting identity, and Arab Nationalism was born.

Perhaps it has been this long association of the Arab with the desert which has now given the matter-of-fact Romans' name of *Arabia* such a romantic ring. For such it certainly had in 1966 when David Holden brought out his book *Farewell to Arabia*. The title struck an immediate chord: like the Near East which had implied some real division at the marches of the settled regions of the Levant, Arabia was gone and the Middle East, stripped of remoteness and mystery, was being swept into the modern world. Since then Arabia has been thought a word of deliberate nostalgia and romance, but it is nonetheless a useful word for the area of desert culture. To escape the charge of romantiscm invited by 'Arabia', some people now use the 'Arabian Peninsula' or just 'the Peninsula'. This does not mean, however, the region falling below a line from the Red Sea to the head of the Gulf, but a good deal of ground in Eastern Jordan, Syria and Western Iraq as well. For, disregarding recent and arbitrary political boundaries, the life-style and cultural traditions found below that line extend far above it until they reach the settled and village culture of the Fertile Crescent and a complex of Turkish, Kurdish, Persian and Mediterranean influences.

The empty lands stretching southwards from the core within the horns of the Fertile Crescent have remained in history austerely Arab. The inaccessibility of the interior kept it thus. An exception is in the south: southern Oman, the Hadramaut and the Yemen, though hemmed into Arabia by the sea, form a distinctive cultural and geographical zone. Other influences from East Africa, the Horn, India and further afield have made their mark on the seafaring and mercantile Arabs of these parts. Rainfall on the high mountains allows them a way of life which, taken with their seaward interests, cuts them off from the rest of Arabia. They do not train hawks.

The area of interest to us, therefore, is roughly a rectangle and may be called desert Arabia. It has its base bordered by the mountains of Yemen (Arabia Felix), by the heights around Wadi Hadramaut in the centre and, on the right, the great sand sea of the Empty Quarter which stretches through Oman and breaks on the salt-flats of the Bainunah, the country between the present day United Arab Emirates and Qatar. On the rectangle's left, from the Yemen north towards Aqabah, run the Hijaz mountains in a line which is picked up again by the hills around Petra on the right bank of the Dead Sea rift and the Jordan Valley. In Jordan, this cultural division between the desert and the sown is neatly drawn by the Hijaz railway, built by the Ottomans to connect the holy city of Medinah with Damascus, and blown up by T. E. Lawrence and his friends in the Arab Revolt. The top line of the rectangle follows that of the Fertile Crescent, from the towns of Homs and Hama below Aleppo, across to Raqqah on the Euphrates, down through Deir al-Zor and along the edge of Mesopotamia to the delta above Kuwait. The Gulf seaboard completes the fourth side.

Within these boundaries, regular agriculture without the help of irrigation from wells and tanks is not possible. The only open water which lasts the year is at Azraq in central Jordan, and many of the pools there have a high degree of salinity.

This area is the 'Chol', or desert, highlands and plains of hard earth intersected by dry watercourses, or wadis, blowing with woody tamarisk and artemisia bushes, sweeps of open country given to dust and in places sand – an anvil on which one learns the value of water. Though in length there is little to choose between Arabia and India, the most distant places have a familiarity which would be impossible in a land more variegated with fields, mountains and rivers. Unless one is in the vein to contemplate the gradual changes in the colours and contours of the desert, to notice a hare's track or plover's nest in the stones, views in Arabia inevitably appear to have an unrelieved sameness. The monotony of the fore and middle ground pulls the eye up to the horizon where, with the heavy bowl of hot sky above, it stays. Not surprisingly, this is how the bedouin navigate, watching and noting the passing bumps and light or dark patches on the skyline, and always neglecting the confusion of tracks and marks under their feet. It can take two or even three years for the tracks of a single car to be worn away by the weather, and so the way is always at the limit of perspective and never closer to hand.

The summer lasts from April to November. It brings sharp contrasts: the black patch of sweat on a goat skin water bag; the white of a stray camel half a mile away in the haze looking as big as a house; and black tents pinned out on a white ground. The dust and the sun destroy colour and the mirage deals with the desert's scarce features. The winter, like the summer nights, is unexpectedly cold. There are more clouds than rain, throwing the world into a camouflage of shadows: browns, ochres, iron reds and, across the flint and basalt plains in the north, blue-blacks and indigoes. The rain, when it does fall, covers the desert floor with a verdigris of a million sharp blades of grass, tiny black irises and other flowers in red, yellow and violet whose names, if I was ever told them, I cannot now remember.

The cold weather too is the time to see wild life, for the summer imposes on those creatures that stay behind to brave it a fleeting owl-like existence to escape the harshness of day. Though now empty by comparison with the descriptions of travellers of forty years ago and more, when there were ostrich, great herds of gazelle, oryx, and partridge in their thousands, the desert is still rich in birds and animals. How many species of lark, piper and wagtail, ravens, gerbils and mice, extraordinary beetles and lizards, hares (inevitably known to bedouin with English as rabbits) and whole families of other beasts, including an occasional wolf, hyena and snake, make up the desert's fauna.

It is hardly strange that the people of such an environment, the bedouin, should consider themselves a class apart. They not only make their lives in the wastes where strangers go in constant fear of getting lost and dying of thirst, but they revel in feats of endurance. In the past, they made a glory and an honour of plunder, and still with reckless demonstrations of generosity and extravagance they scorn the parsimony and prudence of the villager. The bedouin is uneasy in the town and is distrusted. But when he returns to the white emptiness of the desert, he is exhilarated by his release, leaving behind him the farmer and the grocer to tend their tomatoes. He will drink from his camel's milk, subsist on a few handfuls of rice or unleavened bread, but kill a sheep or goat to persuade a passer-by to stay and eat with him.

The bedouin's fugitive way of life has inspired a good deal of romance in the hearts of those who cannot share it with him. In some people the receding image of the camel and rider seen from a passing car stirs thoughts of a bird-like freedom, wings on which to explore the wide silence of the desert. The rider passes on his slow and unknown journey and is a rebuke to the hurried anxiety of our modern ways. He is independent and at ease in a landscape where a puncture or breakdown would be a calamity. We find him by turns inviting and irritating.

Some have visited the desert and found only disappointment of their private hopes. They return dismissive and convinced that others who have travelled there and found it otherwise must have been up to no good. The bedouin, they say, are dishevelled and feckless, by no means what they are in books. An attitude more damaging to the bedouin is indifference. They had some status as a nuisance to the townspeople in the days of raiding when communications depended on the nomads' goodwill and guidance. But now, for the purposes of everyday life, they are no longer an issue and the townsmen can content themselves with the thought that it is now they who set the pace and the bedouin must keep up as best they may.

This relationship between town and desert has been complicated by the development of oil, not only as the currency of progress but also of political persuasion. The Levantine's acute sense of self-interest restrains his atavistic views on the bedouin when he comes to deal with the desert states which produce the oil subsidies. The latter, intending to take a role on the stage of Middle Eastern politics, have readily accepted, at government level, policies and attitudes which bear little relation to their local culture. The old tensions have been recast by a common need to exploit the revenues from oil.

So today, the clothes of the desert are national dress in Kuwait and symbolize affluence and influence. In Saudi Arabia and the lower Gulf, they are worn, with a view to preferment, by men born in Palestine and Egypt. And the bedouin himself? His kind is scattered: men of bedouin stock are to be found running an import-export business, riding with camels in the wastes, training to be fighter pilots or negotiating the price of oil and the stability of the free world's economy in the councils of OPEC. If ever there was such a man as the typical bedouin, he is increasingly rare today. But to own a Mercedes is not to enter the society which made it; and the heritage of the desert is to a greater or lesser extent present in the heart of every Arab.

The exterior variety and change are recent. The time at which the affairs of Arabia turned from confident tradition to a stampede of progress may be put in the thirties and forties. By this time the Al Sa'ud had established its hold over the heartland of Arabia and the prospect of their oil wealth was coming into focus. The Hashemites in Iraq had imposed an administrative structure and some security on that part of the Fertile Crescent. In Transjordan, Glubb Pasha was to eliminate tribal raiding and weld the nomads' allegiance to the central government by the brilliant expedient of putting them in uniform. In the Gulf, air links to India and the search for oil heightened British awareness of the need for security and the 'foreign presence' in those parts was proportionately increased. After the war, these trends accelerated at enormous speed.

The older men of those days would have remembered the Arab Revolt as fairly recent history and know life in the desert as a matter entirely in the hands of God and his harsh elements. Though some of the grandest sheikhs had stone houses in the towns and oases, the vast majority of bedouin lived in tents woven from the black hair of goats. They were nomads in the purest sense. With agriculture, save that of dates, technically impossible, their principal livelihood was raising animals which could survive in the desert. The camel, the best of these, needs a wide range so the bedouin families travelled after their herds, hoovering the thin pile of desert vegetation by the month, and drifting with the season in search of cooler places. In areas with good wells, a family might keep some goats or even sheep which need watering more frequently. But the inner desert only permitted camels, and those who dwelt there felt themselves of higher caste for their more independent and austere existence. Such men would have taken part in tribal raids, poaching livestock from hostile neighbours and pursuing blood feuds. When not on a rare visit to a settlement or toying with the political advances of foreigners, they led a pastoral life not greatly changed since the domestication of the camel.

The children of those times are now only in late middle age, and in those places where development came later, are of course much younger. Some of them have become businessmen and officials, others have clung to their traditions and stayed in the open as nomads. Their possessions have changed, but not uniformly. Some families migrate as of old, on foot or riding, others in brightly painted pickups and even lorries. Others scarcely move. Their stock is watered by tanker or grazed out by some members of the family, while the others earn cash as soldiers and, very popularly, long distance drivers.

The visitor to Arabia is often told that the desert life is over, and in absolute terms of independence and freedom this is so. The tribes are no longer in open and violent competition, but must jockey more subtly for influence and position by infiltrating their members into the machinery of state. If, however, the visitor is impressed by the speed and sureness of this social revolution and asks how the Arabs will fare when their oil is spent, he may be told, as I have been, that they will return to the desert. 'We don't care, we are Arabs!' The question makes impossible assumptions, as does the defiant reply. But the spirit of the rejoinder is unmistakable; its loyalty to the settled life is still imperfect.

To be a bedouin in the seventies, then, is as much a state of mind as one of motion. To describe the frame which houses this mind, rather than constantly hedge with reservations and exceptions to allow for our own extraordinary times, let us take that figure from the forties. He still exists, as I have met him, wandering with his family. He emerges where you do not expect him and connects us in one lifetime with a world which, from the skyscrapers of Jedda or Dubai, seems as distant as the days of the Bible.

A man brought up in the desert is of spare build and seldom above middle height. His face, which is heavily lined by the time he reaches middle age, has sharp cheek bones and a high-bridged nose almost invariably underlined by a moustache. In the south, full beards are common but in Najd and the north, the bedouin, if they wear one, usually shave along the jaw to leave only a small beard on the chin. These striking features are made the more so by the frame of red and white checked or plain white cloth which forms a head-dress. This is made from a large square folded diagonally like a woman's headscarf, and tied in a turban or held in place by a doubled length of rope called an *'agal*. (The root of this word means to couch a camel: the loops of hair cord worn on the head are just the size which you slip over the knee of a kneeling camel to prevent it from rising.) A long shirt reaching to the ankle with a sarong or baggy linen drawers underneath completes the bedouin's dress. The gold-piped cloak or *'aba* is now worn usually by older men or on formal occasions.

His house (there is no separate Arabic word for 'home') is a rectangular tent made from panels of woven goat hair. The front is usually left open and the tents in an encampment stand in parallel lines like cattle with their backs to the wind. The interior is divided into two to protect the privacy of the women and children on one side and the peace of the men gathered in the *majlis* on the other. There the only possessions are some brass coffee pots (some sheikhs have tremendous sets ranging in size from the waist high to the pint) and around these and the hearth are rugs, mats, cushions and saddles making a ring of spare comfort for guests according to the household's means. The rich have tents tens of yards in length; the poor, living deep in the sands, perhaps make do with the shade from a sheet strung between two bushes. All possessions must be easy to carry.

The bedouin move about in family units or several families grouped together. Closely interrelated families make up a *fakhdh* or 'thigh' which is in turn part of an *'ashirah* or tribe; and an association of related tribes constitutes a *qabilah*, a confederation, such as the 'Anaizah of which the Al Sa'ud, the ruling family of Saudi Arabia, are members. Tribes vary enormously in size. The Rualla, another branch of the 'Anaizah, claim well over a hundred thousand souls, while the Beni Hassan are probably less than a tenth of that. Others may be numbered in hundreds. The tribes tended to congregate together in times of danger from raiding in a tribal war and to celebrate feasts. But the great mass migrations of fifty years ago will never return; security and roads in the desert have seen to that. But you can still come across a landscape dotted with tents and imagine dimly the grand scale of the scenes which impressed travellers so much in the past.

That would be from a distance. The romantic aura which the bedouin have attracted obscures the pain and poverty and hardships of which, with extraordinary humour, they appear to make so light. Their own medicine is rudimentary and superstitious but prescribed by those who survive it. Hygiene depends on the habit of moving home across the sterile desert and the meagreness of their diet further debilitates bodies already wracked by the climate. A bad year for rain or an attack from raiders of another tribe could, in the past, reduce a family group to absolute poverty.

This constant insecurity in their lives effectively eliminated the point of thrift or other values so important to the townsman. That is not to say that the bedouin were careless. There are moving stories of tribesmen in the southern sands carrying a small bundle of coffee beans in their pouches, untouched for months at a time, against the need to entertain and honour a guest. But resignation in the harsh circumstances of desert life has bred in them a negligence of material things not essential to a need. Visitors have often noticed the tattered and spartan style of the bedouins' tents as a striking contrast to the dignity and personality of the occupants.

Flamboyant and in turn austere, the character of the bedouin is elusive. Comparisons with other nomadic peoples break down as soon as they are begun. As the people of the heartland of Islam and the architects of its language they cannot be called primitives, but then they have strenuously resisted the civilization which their religion offered them.

Aboriginal Arabians, if noble savages are sought, do exist. The Solubba are not considered by the bedouin as a tribe for the purposes of marriage and so on. They do not keep large herds but make a living as trappers and entertainers (the bedouin disapprove of their own kind playing musical instruments) and always enjoyed a permanent close season from inter-tribal raiding. They do not appear to migrate with the seasons, as do the bedouin, but seem to exist on air in conditions which no bedouin would tolerate. But kindly and inoffensive, perhaps noble, though the Solubba may be, they do not train hawks and are not to our purpose.

The bedouin are later arrivals, warlike but never military. Their history has been one of movement and flux – the drifting of tribes from the comparatively rich regions of the southeast, northwards onto the central highlands and then again in search of better pastures further north to the plains of Transjordan, Syria and Iraq. The present distribution of tribes reflects the

OPPOSITE PAGE *Sticks of wood being gathered by a falconer to build the lunchtime fire over which the morning's kill will be cooked. The hooded falcons will not be fed until evening.*

last great upheavals of the nineteenth century. The weaker tribes who were pushed ahead by these waves gradually took to semi-nomadism, rearing sheep and goats on the edges of the desert as they became excluded from the interior. In a matter of a few generations they lost their bedouin identity. The taut spring of justifying the desert existence by despising the alternative unwound.

The tribal lore of the inner desert has wide horizons because the nomads are always on the move within their zones of seasonal migration and because of the raiding expeditions they used to make into the territories beyond. 'Audah Abu Tayyi, a great sheikh of the Huwaitat, used to lead raids from his encampment in south Jordan to the neighbourhood of Aleppo and into Mesopotamia. But traditions and stories live long among the bedouin and so their lore echoes the events in history of the ancient drift from south to north. So a bedouin from the Shammar, a large tribe in North Arabia whose traditional stamping grounds ranged between Hail and the Euphrates, would know all about the tribes neighbouring his own and, on the basis that a bedouin's enemy's enemy is his friend, the tribes beyond those too. The Beni Sakhr who migrated to central Jordan from the Hijaz in the late eighteenth century still maintain friendly relations with the people of their original homeland five hundred miles to the south. But to most villagers living on the edge of these *diyar*, or tribal grounds, this far-ranging society is a closed book.

A Syrian falconer living near Tartus on the Mediterranean coast, from whom I bought a sparrowhawk, said he knew nothing about the training of long-winged hawks or falcons. He was a cobbler and never had much cause to venture far afield from his home region. The bedouin, he said, motioning with his hand in a distant way towards the east, kept falcons, but he had never had anything to do with them nor did he know their methods.

Later I visited the Sha'alan, the sheikhly family of the Rualla, living at that time near the Iraqi-Jordanian border, and I took with me the Syrian's sparrowhawk. Lawrence, the sheikh, named after T. E. Lawrence, who had entered Damascus with his father on the day he was born, was a fine and successful falconer but could not get over the absurdity of the tiny sparrowhawk, still less its miniature hood. His eyes glazed over with a kindly disbelief when I explained to him how sparrowhawks were flown at quail on the coast. In Aleppo, however, as far again from Lawrence as Tartus, some falcons were spotted in my car outside a cafe by two bedouin over their glasses of tea. They began to discuss the hawks' comparative merits and mentioned Lawrence al Sha'alan as a well-known falconer. Their conversation would have been quite comprehensible and sensible to any bedouin from the other end of Arabia, from Buraimi or Qatar, but to their villager neighbours as meaningless as Greek.

It would be wrong of course to give the impression that desert Arabia is simply an expanse of sameness and monotony in the customs and traditions of its peoples. Those tribes in the north have had richer grazing and so a chance to build up greater communal wealth than, say, those of the Empty Quarter who, until oil, lived at subsistence level. On the eastern edge of the desert, on the shores of the Gulf, some tribes became involved in fishing, pearl diving and owned date gardens, possibly worked by immigrants from Persia. On the northern and western sides, the Arabs were exposed to the Ottomans, traders from Egypt and proconsuls from Europe. As the shore of their desert was settled land and not the sea, their economy was almost entirely based on pastoralism and acting as guides and protection racketeers in the interior. But these differences remain insignificant in comparison with those which the tribes had in common against the settled communities.

Perhaps the most distinguishing characteristic of the nomad is his pride in independence: his environment demands it. In practical terms he is a wanderer, free-wheeling across the face of the desert tied only in winter by the need to find pasture and water in the wake of what little rain may have fallen, and in the summer to wander the country round his tribal wells. His liberty to wander is proportionate to the affluence and power of his tribe, and so to this he and his immediate family give unquestioning loyalty. Without the support and protection of numbers he is defenceless – in the past to marauding raiders from another tribe, and today to economic forces which he cannot face without the mutual insurance of the group. As there is no democratic representation in modern Arabian states, a complicated system of patronage operates and his interests and chances in life are furthered by his relations with his sheikh. A bedouin's first option in marriage is traditionally the daughter of his uncle. So a tribe is a cat's cradle of blood and marital relationships and this further increases its stability and strength.

But while a tribesman's freedom is guaranteed by his submission and assent to the needs of his community, at the personal level he remains fiercely individual. The sheikh of a tribe is not its absolute ruler but a first among equals who holds his position by constantly reaffirming the qualities for which he was originally recognized and therefore by common consent deemed to succeed. The respect which he is given by the tribe must be earned by his reciprocating it, and it is a solemn sight to see a ring of rich and proud bedouin rise up as one to welcome the entry of some grizzled old man walked in off the desert. His age and experience give him position and his shabby clothes and broken teeth are as nothing.

In this tribal world there are no police or social workers to impose a foreign will but, at the same time, unless a tribesman decides to try the odds against him by hiving off as a loner and outlaw, or by seeking the protection of another sheikh as a client, there is no escape from the tight contractual relationships which bind him to his community. The bedouin's courtesy and public restraint therefore come of long practice in communal living. Privacy, as we understand it, is a thing unknown, but discretion is at a high premium and it is disconcerting after a long period with a family or tribe to realize how little one knows of their inner thoughts and concerns.

Honour, defined as hearsay, and 'face' are almost matters of fetish and preclude the open discussion of others' faults – at least in front of strangers. On the contrary, the bedouin are avid for stories which reflect well and can unsettle a green visitor by heaping on him compliments and innuendoes full of delicate flattery. During the interminable songs and sagas recited at feasts and weddings, if a man hears the names of his father or grandfather mentioned, he fires a clip from his gun to demonstrate his allegiance to the family honour, and thus of course increase his own. On first going into the desert, the foreigner probably finds that the most striking aspect of this individualism is the hospitality and generosity by which the nomads seek to express it.

For every story of the sheikh who called in the poor of the neighbourhood to share his meal, there is another of the one who left food out for the howling wolves, declaring that no one at his camp should express a need unanswered. A hawk which I was once given by Lawrence al Sha'alan had been accepted, I was aghast to discover later, as a present from a member of the tribe who had trapped it. Lawrence had given him in return a new Toyota jeep and six hundred dinars – at today's rate about £850. A bedouin precept runs 'If you strike, give pain; if you feed (a guest), give full; and if you give, pour riches.'

But if life in the desert is simple in material things, and attitudes, born of age-long self-confidence, tend to be abrupt and straightforward, this is perhaps because the life is so circumscribed. The problems of the bedouin's day-to-day existence were solved in early history, and in the unchanging environment of the desert there was no great urge for invention or progress. So the bedouin have evolved as survivors and not innovators and those not content with this have left the desert. Tradition with hoops of steel binds together the society which is left. And it must be said that the hand to mouth introspection which developed has made but a very brittle protection against the enormous forces of change which are now attacking the desert and threatening its delicate ecology – the plants, animals, as well as the human element.

Opposite Page *Sheikh Zaid bin Sultan carrying a haggard peregrine. Sheikh Zaid, now president of the UAE, was governor of Buraimi when this picture was taken.*

Page 38, above *A bedouin with his saluki at Hamra Sands on the edge of the Empty Quarter.*
Below *Buraimi oasis in 1949. These Barasti huts of palm fronds and matting have been replaced by a modern Hilton hotel.*

Page 39 *Sheikh Zaid's falconers in 1948–49. The hawks at top left and bottom right are haggard peregrines, the other two are passagers. Muhammad bin Fadil, kneeling on the left, was still an active falconer in 1977.*

Page 40 *A tribesman carrying a passage peregrine. The hawk's full crop shows it has been well fed. The feathers in the rifle barrel help protect it from sand.*

The Arabian attitude to hunting is a good example of this inflexibility. Firearms only reached the desert in any numbers late in the last century and efficient ones much later than that. Before then, the only way to catch birds to eat was to use snares or hawks, and animals like hare and gazelle could only be killed with spears and salukis. Archery was once known in Arabia but given up long ago.*

Hunting was a strictly pragmatic business. In a staple diet of thin milk products, bread, dates and perhaps rice, meat was a rare treat, especially if it came from killing off valuable stock from the herds. So the bonus of a meal of game was not risked lightly. The natural law of all predators was carefully observed: the food value of the quarry had to outweigh the effort expended in catching it. Thus, the delight of 'fatter' races for flying merlins at skylarks, or sparrowhawks at blackbirds, simply for the excitement of the flight, would strike the bedouin as manifestly absurd. You can't eat it, so why catch it?†

As the hawk must also be kept alive with a meal from its kill, the desert falconer had to set his hawk at prey large enough to provide for all interested parties, and this requirement led to the growth of some extraordinary beliefs about what a hawk will or will not take. I have often been with falconers in the desert who will catch a lizard or jerboa to feed up a hungry saker and indeed this hawk does kill such things for herself in the wild. But the same men have told me that hawks have no taste to kill cream-coloured coursers, a fairly common bird in the desert. The subconscious explanation must lie in the fact that a dead courser which has been through the hands of a hungry hawk leaves precious little for the falconer. The same is true of the raven, and my stories of rook hawking in England and country people's liking for rook pie in the spring have been regarded as further proof to my Arab hosts of the utter depravity of the occidental.

The Arabian Muslim's reservations about dealings with dogs (he won't have any), a category from which the graceful saluki is a singular and honoured exception, eliminate the possibility of using dogs to scent out or point quarry. (They also eliminated, by the way, a naïve but well-meant

*Doughty in 1875 reckoned that it had been dead two centuries and records in his *Travels in Arabia Deserta* a conversation with a Sharari tribesman in which the latter tells him of occasionally finding in the mountains arrow heads 'like little spears, but the wooden shafts long perished'. The tribesman regarded them as very strange and identified them with the period of the crude wall-carvings he had seen in caves showing men with bows and arrows wearing long hats, a bizarre fashion to him and out of time.

†Small quarry can in fact make up a pot, as the Syrians know with their sparrowhawks, but only where the game is sufficiently plentiful – conditions which the desert does not supply.

suggestion a few years ago that a number of Saudi Arabians be selected by their police force to attend a British police course for dog handlers.) In the West, a hawk may be got into a position of advantage high in the sky over a dog on point which then flushes quarry which in level flight would be too fast for the hawk. The hawk trained to do this is said to 'wait on' and this is the tactic used in hawking game such as grouse or partridge. The waiting on flight, however, is not known to the bedouin, nor is it available to them as it is best used with dogs.* Their hawks, therefore, must be flown at quarry which they can fly down, and this, taken with the need for cost effectiveness outlined above, limits their scope to three items: the Macqueen's bustard (the houbara), the stone curlew and the desert hare (see Quarry).

These three, but especially the first, are chased, killed and eaten with a relentless passion. But since the First World War and the increasing availability of shot guns and automatics, their numbers have steadily declined. So have those of the gazelle and ibex; the oryx is now virtually extinct and the ostrich completely so. Hunting in Arabia is not a pastime as we might at first understand it. The anxious excitement of battle and raiding but without its element of fear: this is part of the magnetism which hunting has over the bedouin.

*I have often wondered whether a waiting on flight could not be achieved with pointers at sand grouse which are usually too fast for a trained hawk in level flight. I have not been able to try the experiment, but I suspect that the dry ground of the desert would not hold enough scent for the dogs to operate successfully.

THE ARAB FALCONER'S BIRDS

'Not only fine feathers make fine birds . . .' Aesop.

IN ENGLISH THERE IS OFTEN CONFUSION ON THE DIFFERENCE BETWEEN A hawk and a falcon: the words appear to be used interchangeably and without explanation. The falconer's way of talking about his birds sounds complicated today when it is used side by side with the language of ornithology, but it made its own sense in the days when falconry was still a popular sport. Then everybody understood it as a part of his general knowledge. The falconer, however, still needs to name things and make distinctions which would be scarcely noticeable to the layman of today and so he keeps up the old terminology. In Arabic there are similar problems. The bedouin's keen observation of nature is at a disadvantage with birds. Most of them only pass through the desert in winter and he is therefore unable to watch their full life cycle. Consequently, he misunderstands some issues of classification and, with hawks, mistakes their sexes. But the way he judges and names the different species of hawk is based on a sharp eye for their various personalities and characteristics.

The birds found useful for falconry are divided into two types in ornithological language: the falconidae, which comprise those dark-eyed, tooth-beaked falcons which have a natural predisposition to attack from above using the 'stoop', or powered dive, to close with their prey; and the accipitridae, which are the true hawks (i.e. not falcons). The latter include the goshawk and sparrowhawk, yellow-eyed birds of more enclosed country which catch their quarry by 'binding to', or grabbing, after a short and level chase much in the manner of the greyhound after the hare. Falconers call these two types 'longwinged' and 'shortwinged' as the wings of the falconidae are much longer relative to the tail than those of the accipitridae.

Unfortunately this reasonable distinction is complicated by the falconer's custom of reserving the term 'falcon' only for the female of any of the longwinged species, and using the word 'tiercel' for the male; and further by the habit of using 'falcon' only of the female peregrine, counted by tradition as the senior bird of those trained. So a saker is not a saker falcon, but simply a saker and her brother is a sakret. Male lanners are lannerets and male gyrs are jerkins. In a room full of sakers you could say there were so many falcons and the rest males. But at the same time if a purist came in and said 'Look, there's a falcon' you would know he had spotted a stray female peregrine. This inhibits the falconer from using 'falcon' except in that special sense. He feels easier and surer with the readier term 'hawk' and in deference to the female's superior powers generally refers to a hawk in the feminine. I shall keep to this convention though it may seem perverse as the birds trained by Arab falconers are, without exception, from those species known to the world as falconidae or falcons and to their desert masters as males. The Arabs simply call their hawks 'birds' (*tuyur*) and those who use the word found in the average dictionary, *saqr*, are at once identifiable as strangers and townsmen.

Falconers further classify their hawks according to the age at which they were trapped or taken. In the West there are three divisions. The eyass is a bird taken from the nest just before she can fly; the passage hawk, or passager, is one trapped in juvenile plumage, during her first autumn migration; and the haggard is a bird caught in adult feathers, i.e. one or more years old. An eyass cannot become a passager nor yet by extension a haggard, just because she moults her feathers: the type is determined for good at the moment of capture. The eyass or passager which has moulted is called intermewed and the haggard simply remains haggard.

The Arabs' scheme is the same, with the important exception that they make no common use of eyasses, which are bred, after all, in lands distant from their own. The passager is called *farkh*, meaning young bird, and the haggard, *qurnas*, a word whose root sense in Arabic is 'hunting'. An intermewed passager is *qurnas al bait* (of the house), while the haggard is strictly *qurnas al hawa* (of the air).

The Arabs' disregard of the eyass is based largely on two considerations. Primarily, eyasses have to be taken in the middle of the summer when the desert falconer is locked in a close season imposed by intense heat, scarcity of

OPPOSITE PAGE *Passage peregrine*

hawk food and even greater difficulty in finding quarry: not for him any glorious Twelfth* with a team of grouse hawks, nor yet any pleasant evening walks after summer rabbits with an eyass goshawk. A few hares would be the sum of his bag, even if he could bring his hawk into flying condition in such adverse circumstances. A natural consequence of this is that the Arab's interest in hawking almost seems to be governed by the seasons, latent enthusiasm waxing to fanaticism as the weather cools.

Secondly, and one must allow that this is perhaps due to their lack of experience in dealing with eyasses, Arab falconers are as one in deeming the eyass a dim pupil. One falconer put his dislike to me succinctly, saying, 'the trouble with eyasses is that they don't leave the falconer, they're forever hanging round waiting for the lure – it's the only thing they know how to catch'. If they do achieve some kills, they never in Arab estimation reach the standards of flight shown as a matter of course by the passager and haggard who have both been living by their wits and skills until their capture.

Many Arabs' complete ignorance of eyasses is illustrated by the fact that many are sold to them at passagers' prices by dealers from Asia, even though they show the sure signs of the eyass – screaming for food and extreme tameness when hooded. The Arabs distrust these qualities but seldom seem to guess their cause.

The haggard and the passage hawk, though they share this qualification of experience and savoir-faire, differ considerably from each other in temperament. The older bird has probably already established her likes and dislikes in the matter of quarry and, more importantly, may have decided after nasty encounters in her youth that some of the prey for which her falconer intends her are a game not worth the candle. Thus a haggard which, for example, has no scruples in taking on a houbara will be a deadly ally; but the bedouin say that these are scarce. Most haggards, they claim, have lost their hot blood and are unreliable. In addition to this basic question of whether they will go, haggards are by no means simple to train and at best their obedience is uncertain. Their impressive dignity and self-possession come of a reserve which is never wholly laid aside and can, in the event of the hawk being lost, quickly take over to make her wild and unapproachable even an hour or so after she has left the falconer. But this dignity and poise does make the haggard attractive to some older falconers. After a lifetime's training of docile passagers, the more mercurial haggard is a greater challenge and its mastery is a suitable gesture of lifemanship towards the young men who perhaps have neither the time nor patience to attempt it. Another general

*August 12th is the start of the grouse-shooting season in Britain.

point in the older birds' favour, however, is that if captured in good health, they prove themselves survivors and have great resilience.

The passager is for the Arab the optimum. She combines the tractability of youth with its capacity for adventure and carelessness of danger. Hatched in April, fledged and flown by the end of July, she will have only had between three and five months of life in the wild by the time she is trapped in the autumn, and usually scarcely this period as the young birds tend to begin their migration ahead of their parents. During this time the passager has been living off the fat of the land, firstly on the young and feckless children of the summer on which she learns her trade, then on the dense groupings of prey on the road south. She is quickly trained for she does not have that mature wildness which alienates the haggard, and can be relied on to a much greater degree to remember her lessons when flown at quarry.

When taken off her kill, she is less prone to the fits of outraged sulking which can overtake her elders, and yet has none of the snatching and clambering bad manners which mar so many eyasses. These on a kill, or even when feeding on the fist, often behave just like small boys who do not want their work to be seen by a classmate: crowding their wings around their food and scratching and glowering at any intrusion. Amenable and competent, the passager is *victor ludorum* in the matter of age.

With regard to the question of sex, it should be noted that male hawks are approximately one third smaller than their sisters, hence their name in English, tiercel, meaning one third. In Arabic a similar convention applies. For example in the case of the sakeret the name is *mathluth* (*thulth* meaning a third), though the Arabs think of the bird as female. Most tiercels take stone curlew but only exceptional ones are capable of holding down a hare or houbara. As quarry is never easy to find in Arabia and falconers therefore require their hawks to be able to take on whatever they can find, the tiercel is not popular: he does not fill the pot and few bother to keep him just to specialize in stone curlew. There are exceptional cases. A particularly beautiful or powerful male, especially if he is a peregrine, may win a place in the *majlis*. But by and large the Arab falconer only trains females; they are tougher and therefore easier to keep and do not let him down in the field.

As explained in the preceding chapter, the Arabs are only intent on three species of quarry – the houbara, the hare and the stone curlew – and their hawks are therefore exclusively those which will most efficiently bring these creatures to book: the saker, the peregrine and the lanner.

There are other birds of prey which could be used to catch the bedouin's quarry but the desert Arabs do not consider them. Some of the reasons are technical. The smaller eagles used by some of the tribesmen in central Asia would take too much feeding from the bedouin's precious larder, and may not have the persistence to see through to a kill some of the very long flights which the open desert makes possible. The same may be true of the goshawk, more at home with short dashes than the persistent chase, but in any case it is not often found in the desert where the Arabs trap their birds. Though these species were used by others in earlier times (see Appendix), the bedouin do not regard them as worthy and are not even ready to experiment with them. The force of tradition is strong. The hawks they do use combine the size, great speed and perseverance which make them ideally suited to the bedouin's needs and are available at the beginning of the hunting season.

Most difficult to assess or explain are the bedouin's emotional feelings about his hawks. No falconer, from East or West, has the kind of relationship with his hawk which would be possible with a cat or dog. The bond is contractual and, if it becomes an affection of sorts on the part of the human, this is because increasing success leads to increasing trust. Nonetheless, a trained hawk is a beautiful and composed creature and in Europe longwings (falconidae) have always been thought more so than shortwings (accipitridae). Though the bedouin do not make this comparison in practice as they do not train the latter, the longwing's grace is not lost on them and they admire her for it. But I think this is a very secondary thought: the bedouin are not self-conscious aesthetes. In the West, men become falconers because they want to see hawks fly and their traditions have developed to create the most taxing flights possible. In Arabia the emphasis has been different. Hawks are trained because they can catch food and a good hawk is a brave one which does not shirk her task.

A memory which may illustrate this pragmatism is of a moonlit night in the desert. We had a wonderful view across a plain and I asked one of my companions if he thought the moonlight beautiful. Rather than take flight in any rhetoric or romance, he simply said that in the old days the Arabs thought the moon was a bane because they could not go on raids when it shone. Nowadays who cared whether it shone or not – it did not matter anymore.

Of the Arab falconer's birds the saker (*Falco cherrug cherrug*) comes first as the most widely kept hawk and one best suited to the bedouin life. The largest of the type sometimes known as the desert falcons, the saker is a robust all-rounder: her tough constitution enables her to thrive on the thin diet of goat or mutton which, on off days, may well be all her falconer can get for her. Her soft and long feathers, almost mop-like in repose, withstand the inevitable knocks and scrapes which they suffer when the hawk is kept on a block and not on a higher perch as in the West, and, more seriously, when she closes with quarry on the ground. Her long and powerful legs are also a great asset when taking her prey on the ground and this she seems quite happy to do, unlike her cousin the peregrine, avowedly a hawk of the air.

When first meeting a saker, and she is the one most commonly seen in photographs of the Arab with his hawk, her most striking feature is perhaps her eyes which appear extraordinarily large and fixing. The bedouin believe that the saker has more powerful eyesight than the peregrine, and as in Arab hawking the quarry must be walked up or flushed by the falconers, this is another point in her favour. A demonstration of this apparent fact may be the saker's readiness, in the wild state, to find and kill ground quarry, while the peregrine is more stimulated by the wing movement of birds, a more visible prey. (There is an idea that the saker's superior eyesight comes of more densely packed cones in the fovea – the section of the retina where the light is most concentrated.)

Sakers do not breed in significant numbers in Arabia, but in the regions forming a belt across eastern Europe, the Balkans, Turkey, southern Russia and central Asia. Less fastidious than other hawks, their natural diet covers a broad spectrum from quite small animals, such as lizards and mice, to large birds like the houbara. Sakers differ enormously in size, small sakrets weigh as little as 1 lb 2 ozs (510 g) and large females up to 2 lb 6 ozs (1 kg) or more. They also vary widely in shade, from deep chocolate browns to pale buffs, their heads in some cases appearing almost white.

Their feet, again reflecting their catholic tastes, are, in proportion to their bodies, small and muscular when compared with those of bird-killing peregrines, merlins and sparrowhawks. A greenish, duck-egg blue in young birds, graduating to a pale yellow in adults, in captivity their feet are prone to disorders like corns and inflammatory conditions such as bumble foot. This is thought by some to be a result of the saker's somewhat lethargic behaviour when tame and at home on the block, causing a slight compression and decomposition in the scales on the ball of the foot and the tissues beneath them which renders the foot less protected against infection. (A hawk's claw has a concave underside which can harbour all sorts of bacteria. The hawk may inject herself with these germs if she punctures or scratches the ball of her foot with an overgrown claw when snatching at her food or refusing to release it.)

In temperament the saker has much in common with the cat: cunning, deceitful, ferocious in the fray, given to laziness, but tameness itself when in the vein. A saker will often lie down for better balance between driver and passenger on the bumping front seat of the landrover or jeep. At the camp, her block may well be in the circle of men around the fire and she will sit there, bare-headed but relaxed, one foot drawn up into her fluffed up breast feathers – an intimacy which would be anathema to a peregrine. Yet when setting about a hare or landed houbara, she is savage and courageous, riding the somersaults of her frantic prey with a grim tenacity. The Arabs consider the saker to be more intelligent than other hawks and it is true that she does appear to think more carefully in deciding her tactics than does the prima donna peregrine who relies more on her speed. Short-cuts, diversions to exploit dead ground, feints and false attacks to break up a flock are all characteristic tricks of the saker which make her such a rewarding hawk to fly.

To achieve these desirable frames of mind, the falconer matches the saker mood for mood. Her obstinate wildness when first captured is met with hardness and a resolute denial of food, her tendency to slackness in the field with sharp dieting, but endless care and attention to cultivate her ready humour. The saker is thus well adapted to the style and methods of training she must face, and, though she lacks some of the finer qualities of the peregrine, she is given full credit for her worth.

Hurr, noble or free, is an emotive word for the Arab and he applies it to five things which for him are beyond price, though they may, he reckons, be obtained with presents. They are the pure bred Arabian horse, the saluki, the thoroughbred riding camel, the pure blooded (human) Arab and the female saker. The saker will be *hurr* if she is *kamil* or 'perfect', i.e. measuring the span of a man's arm, elbow to finger tip, from her head to the tip of her tail. Birds of medium size may be called *mathluth* and smaller ones, *garmush*. The generic name for the species is *saqr* from which of course comes our own word, saker, borrowed by the Crusaders when they first met this hawk.

Next comes the peregrine (*Falco peregrinus*), a specialist. She gives an immediate impression of being wholly designed to kill birds in flight and this is her true vocation. Her long and stiff primaries cut effortlessly into the head wind and her short tail betrays her trust in speed and not manoeuvrability; her long and almost spindly toes are for pinioning and not paralysing. She usually attacks by stooping from a height, dealing her prey an often fatal blow as she passes. She then lands on it and at once breaks its neck with a powerful wrench of her toothed beak. But if she scores through virtuosity, it is at the expense of versatility – the saker's forte. Her feathers are easily broken in disputes on the ground and, not being designed for dry hot climates, they quickly lose their gloss and bloom, making her appear something of an uncomfortable stranger in the desert scene. She requires careful feeding on natural food, being unable to tolerate the meaner staples on which the saker can keep fit.

The peregrine is an increasingly rare bird. In the lower Gulf, where twenty years ago the peregrine was the commonest falconer's bird, she is now at a high premium, and in the north and Mesopotamia trappers now say that for every one they see, they can find ten or twelve sakers. As a consequence, peregrines command ever high prices, eight thousand dollars being a good bargain for a passage female in 1977. They are not found far from the water where they can expect to meet up with their favourite quarry and are thus called *shahin bahri*: *bahr* being the Arabic for sea.

In medieval England she was called the 'falcon gentle', and this reflected not only her elite social status but her temper, which, in comparison with that of the saker, is aloof and reserved. This introversion makes her an uneasy guest in the desert household where privacy is unknown and the habits born of long dealings with sakers are sometimes applied to the peregrine. These conditions do not bring out the best in a passage hawk, but in the haggard they more often than not neutralize whatever potential she had when trapped. The good falconer, however, is aware of these difficulties and does his best to mitigate their effect on the hawk. He keeps her hooded when she is likely to be disturbed, which is most of the time, and does not subject her to the stiff course in 'manning', or taming, which is *de rigueur* with the saker. He prefers not to fly her at hare but keeps her for those flights where she can show off her incomparable and superlative powers.

I commented once to an old and brilliant falconer how impressed I had been with the gentleness of a young sheikh's handling of his peregrine. The hawk had had a long and hard flight at a hare but had lost the race to the latter's bolt hole by a snatch of her claws. When we arrived, she was sitting by the entrance to the hole, panting in the strong sunshine. The sheikh, realizing that her blood was up in a high temper, lay down flat to crawl towards her and not frighten her by stooping over her. 'That was well,' noted the old man, 'he could be forgiven the flight at the hare for he was young. But he should have seen to the hawk's comfort before picking her up. I once had a similar case: my hawk was hot, so before taking hold of her, I took off my head cloth and held it out over her. This created some shade and at the same time a tunnel to catch some breeze to cool her. She then stepped quietly onto my fist when she was ready.' As an artist in the management of hawks, it never occurred to him that the hawk would not be tame enough to accept such treatment. He, after all, had trained her.

Though significantly smaller than the saker in size, though not necessarily in weight (most peregrines fly at about 2 lb 2 ozs, 900 g), she is always the faster and if brought to the field on good form, does not show the lack-lustre indifference which sometimes spoils the saker's day.

The tiercel, in Arabic *tiba* or *qusiyyah,* is sometimes kept on when a sakret of similar size would have been let go. A carefully managed tiercel will sometimes kill a houbara, but at stone curlew he is adept. He is nicknamed *samm al-kairuwan*, stone curlew's poison.

Such are the peregrine's qualities for which she is acknowledged as the supreme performer. But though her rarity further increases her value, she does not in general command the affection which the *hurr* has won in the hearts of the nomads.

Like the peregrine, the lanner (*Falco biarmicus*) is known as *shahin,* but in full *shahin wakri*: *wakr* meaning nest or eyrie. She does indeed resemble the peregrine when compared with a saker, but in the bedouin view is altogether from a cadet branch of the nobler family. Smaller and lighter armed than the *bahri,* it is an exceptional lanner which will take a houbara. She is quite often flown at hares but must be harshly dieted to get her to take them – an indication of the bedouin's lower esteem for her.

The lanner is easy to train and placid when made. Being another of the 'desert hawks', she can accept a coarser diet than the peregrine, but like the saker she has a corresponding tendency to lethargy. She has a sharp intelligence and a very attractive head to hold it. Her pale rufous and cream nape is highly regarded and usually accounts for her price if she makes one. In these more monied days, the lanner is often disregarded as more useful hawks are brought in from far and wide and there is the cash to buy them. But she was more commonly pressed into service when hawk trapping was not the big business it is now and falconers trapped for themselves, having to make do with whatever they could find. A keen falconer, however, will still try out the second-class *wakri* by way of a change, and if she achieves a success she quickly becomes by virtue of her hardiness and tameness quite a favourite with the family.

The red shahin (*Falco pelegrinoides babylonicus*) is primarily a bird of Asia and the Indian subcontinent, but some are trapped from time to time in Arabia. Though better proportioned than the lanner, this hawk suffers from the same disadvantages of small size and lightness which disqualify her from success in most flights, though at stone curlew she is in her element.

I once bought a female for twenty-five Syrian liras, about £3.50. She was the size of a peregrine tiercel and very beautiful. Lawrence al Sha'alan, who saw her a few days later, offered me a dinar for every lira I had paid. He said he had had another like her who had killed houbara and was sure mine would do the same. There were twelve lira in a dinar, but I refused.

The red shahin is usually confused with the lanner because of her red head and neck, or, because of her very large feet, with a peregrine tiercel and so she is known as *wakri* or *tiba'*. Very large females, which may weigh 1 lb 14 ozs (950 g), are sometimes passed off as small female peregrines and, if they have spirit, are no disappointment to their buyers.

Though the gyr (*Falco rusticolus*) is by no means a hawk traditionally trained in Arabia, its recent arrival at the hands of oil companies, banks and large corporations who give them to rulers interested in hawking, forces her inclusion along with the red shahin as an occasional visitor. In the Gulf, some falconers say that they have met similar birds in Pakistan and call it *sunqur* or *shunqar*, the Persian word for the gyr. Though it is not possible to exclude the northern races of gyr, as some may have been brought south in the past as presents like their modern successors, it is probable that the *sunqur* is traditionally the Altai gyr which is much browner in colour and loosely resembles an enormous saker. In fact Arabs who do not know the Altai gyr as the *sunqr*, or are not observant enough to notice the arrow headed bars on the lower edges of her breast which no saker has, call the gyr simply *hurr*.

The gyr is widely known for her immense size and glamorous white plumage and these are just the qualities with which she has dazzled Arab falconers, among whom size and pale colour are popularly counted as marks for preference in choosing hawks. If a blonde saker is admiringly handed round the *majlis* as a proud and rare possession, imagine the stir caused by the entry of the first Arctic gyr, white as her native environment and of a size to kill a saker, one in each foot. There, however, the success story ends as the bedouin have quickly remarked that the gyr easily sickens and dies in their climate and, if brought by good chance to flying condition and the field, has a flippant and nonchalant attitude to the business in hand.

A further reason which may put her at disadvantage with the Arab may be the fact that most gyrs sent out to Arabia have been taken as eyasses and not yet entered to quarry. With such hawks the Arab falconers have little or no experience and can therefore scarcely be expected to bring out the best in them. The fact remains, however, that their weight and delicate health make them hard birds to fly well in Arabia and they generally remain *tuyur al majlis* ('drawing-room hawks') condemned to lend elegance and exotic cachet to their master's entourage for the time they survive.

OPPOSITE PAGE *Saker*

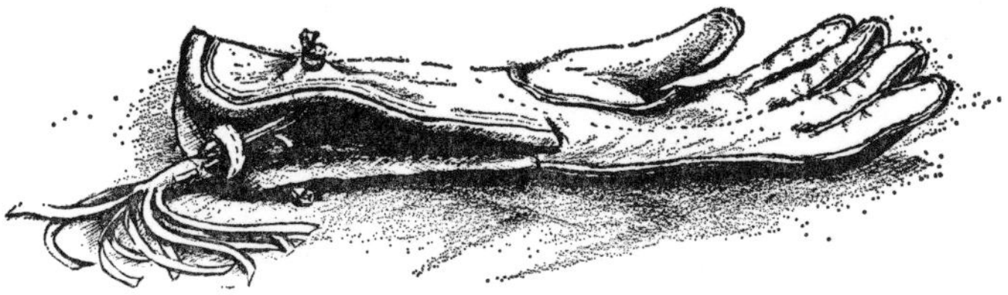

TRAPPING

'In the ways hast thou sat for them, like an Arab in the desert.'
Jeremiah.

HAWKS ARE CAUGHT ON MIGRATION IN THE AUTUMN. AT THIS TIME Arabia provides two major routes for all birds moving from the vast catchment area extending from south-eastern Europe to Russia and central Asia. The first of these forms a band from the Bosphorus to the eastern Mediterranean coast, down to the line of the beginning of the Great Rift Valley in Syria and the Lebanon, through Transjordan to the Hijaz mountains and the Red Sea. The second draws from the Caucasus and crosses the mountains of Kurdistan down to the plains of Mesopotamia and the Gulf (see map page 8). From September until mid-November, the ground covered by these two courses is a paradise for the bird watcher. And as the millions, from pelicans to linnets, struggle south, so with them come the birds of prey – eagles, buzzards, falcons, hawks and harriers – in their thousands, pushing on the able-bodied and killing the stragglers – the drama of natural selection.

The techniques which the Arabs use for trapping their hawks are simple and may even appear apocryphal, but to the wild hawk their tricks are less obvious. But that is not always the case. It is unlikely that many of the migrating hawks have not seen humans before and there are many stories suggesting that trappers, to catch intelligent hawks, have had to disguise their occupation and not their species. At all events, the development of the trappers' methods has been the proof of their success. They comprise one or a combination of the following methods and ruses.

THE PIGEON A light frame, into which have been sewn about a dozen upstanding open slip nooses, is harnessed onto the back of a pigeon. If the hawk kills the pigeon, or even just grazes it with a stoop, her long toes will become entangled in the nooses. Though she is able to fly off with the pigeon dangling from her feet, she will find this impossibly awkward and alight at no great distance in frustration and anger. In this frame of mind it will be possible to run in and throw a cloak or head cloth over her.

THE DECOY Hawks, being inherently jealous and always on the watch for a simple meal, may well be tempted to rob a smaller hawk of its meal. A small sakret or lanneret is half seeled (i.e. its eyelids are half stitched up, see p. 62) and a bundle of feathers liberally reinforced with nooses is tied to the hawk's jesses. When thrown off, a seeled hawk will usually fly in confusion upwards and in circles, trying to avoid flying into obstacles, then probably flutter down to the ground and a crash landing. Likely as not, this will be too much for the intended hawk which thinks the decoy is carrying a kill and becomes ensnared in her attempts to rob it. The seeled hawk is known as *niqil*, a term of affectionate derision often used of any diminutive hawk.

THE NET Predictably, though not to the hawk, the net is a fine, wide mesh net some three feet high and five wide, which is lightly rigged up in the form of a lean-to. The two upper corners are supported by two long reeds or fine stems and the two lower ones buried in the sand or held down with gravel. Beneath and up-wind of this, a pigeon squab or a jerboa is pegged out as a bait. A hawk coming in to take the bait will do so up-wind and either crash into and be enveloped by the net in the rush of its arrival or, if unusually she comes down the wind (as will some lanners), she will be carried into it by the momentum of her swoop. The hawk has only two solutions. Either she must land inquisitively beside the trap and walk onto the bait taking care to depart on the open side – a course of events so rare as to be discounted – or knock the net down with a preliminary stoop. Hawks have been seen to do this many times but it is a risky business. Her problem is one of perspective: almost certainly she sees the net, but misjudges its position.

THE HIDE This method seems so extraordinary that I would have put it down to romance had I not met men who have succeeded in it; inasmuch as others have corroborated that they have left the trapper in the hide with nothing but a pigeon and he has returned with a fresh hawk. A small hut or hide is built with matted palm fronds and brushwood above a scooped-out hollow in the sand, usually on an island in the Gulf or on a sand-bar.

The trapper watches from inside with a live pigeon on a line. When he sees a hawk, he pushes out the pigeon into the wind where it flutters on the end of the string. The hawk, God willing, kills it and begins to deplume it and then feed. She will do this facing into the wind so that her own feathers are not ruffled and those plucked from the pigeon are blown clean away. And she will therefore be facing away from the hide. While this is going on, the trapper slowly, slowly, slowly draws in the line and with it the pigeon and feeding hawk, along the smooth sand. An arm, the while, has been lying in the sand, and when the hawk is well within reach and looking the other way the hand grabs her by the legs.

A variation of this procedure is to bury a man in the sand with his head camouflaged with brush. One hand protrudes and in it is a live pigeon held firmly by the legs. Should a hawk kill this, it would be not a great matter in the ensuing pushing and pulling to extend the grasp from the pigeon to the legs of the hawk.

Three conditions only are necessary to the success of these operations: patience, nerve and plenty of hawks.

In practice, the days may pass in nothing but frustration. I have been in the desert with trappers for fifteen days at a stretch, seen very few hawks and caught none of them. The bedouin believe that the migrating hawks pass through in waves. The intervals between them can be empty and disheartening. But the times one remembers are an extract of all that was best. On some occasions the early morning desert has seemed almost littered with sleepy birds of prey. With luck the trapper's day may follow along these lines:

An encampment in south-eastern Jordan before dawn on an autumn morning – a cold and silent sight. The trappers hurry to pack the car or pickup with their supplies and cages of pigeons. They hang the noose harnesses in their customary place on the rear mirror. No time can be lost as they must reach the open ground well before light. A few biscuits or dates may be handed round as they set off, to work up an appetite for cigarettes: a time of great excitement. They draw away from the village and, as streaks of light stain the eastern horizon, they make the wide gravelly levels and mud flats where hawks are found roosting during the migration.

A small black form can just be made out behind a shrub away on the left in the gloom. The car pulls round to drive past at a tangent but it is only an *'aqab*, a small eagle or buzzard. On, without the headlights now as the gloom clears fast, and across the undulations of a narrow watercourse, corrugated by the rare flash floods in the spring. The snapping shrubs in the water bed throw up a strong perfume as they brush under the tyres – a good chance for a puncture – and out again onto the glassy noises of the gravel. Twenty minutes later, another form, but it spreads its wings and swims off – *abu haqub*, a hen harrier.

The other trappers have set up a base camp in the desert with a couple of tents and this appears in the middle distance on the edge of a hollow. All the cars are gone, to return at about half past nine when the sun has warmed the air into thermals and the hawks have taken to soaring and become indifferent to food as they move on in their journey. Until this time, however, they are lethargic, staying where they have slept long after other birds have taken to wing; a self-indulgent trait which falconers take into account when they put out their hawks in the morning to 'weather' on their blocks before flying them. Some trained hawks, indeed, will take themselves off to some perch to preen and weather for themselves if they are taken out to the field too early in the day, or without being allowed to weather at home.

A hawk, on the right, takes off from behind a small bush as the car passes – a *shahin*, a small one, probably a tiercel, but worth a try. The car slows as the hawk banks to look (they do not often bolt away in a straight line) and the man driving is handed a dressed pigeon. He checks the loops which hold the ends of the harness to the pigeon's thighs, and with a steady sweep of the arm throws it away from the car, always on the off-side from the hawk to give the impression that it was flushed by the car. It flies for twenty yards and lands, hampered by the harness. The car keeps going gently and stops after a quarter of a mile. The hawk sees the pigeon and makes for it with those rapid, pipping wing beats of the tiercel in action. But he does not come low enough, passing over the pigeon wide and high. After the cold night, hawks are usually hungry but not always quite enough. This may be a case in point. He banks again and makes another approach. The pigeon starts running for a bush, struggling with its burden but perhaps sensing that its only safety lies in not taking off. No stoop but a close swing over the bush. Once, twice again then at the top of his climb, the tiercel holds his wings out and glides, a half turn and then slowly away on the wind.

The pigeon must be recaptured and the hawk tried again, perhaps with a more lively pigeon who may excite him more. As the car pulls up and the doors open, the pigeon, in a state of thorough alarm, makes for safety and the underneath of the car. This always precipitates a fair amount of slapstick diving and falling about which at least breaks the earlier tension but makes the pigeon even more jumpy until someone gets the special length of wire with a crooked end which will bring it in by a noose.* Then off quickly.

*The wire is called *al-mikhtaf* – a good example of the richness of bedouin Arabic. The wire has no other purpose, so it has its own name.

More cigarettes en route, then pulling up on a rise they get out to scan the ground before them with field glasses and search for the tiercel. In the distance a car is stationary, they may as well ask for news – a reflex reaction among the bedouin on seeing a chance of company. Approaching, they recognize the Landrover for a friend's but suddenly see why it has stopped and halt also. Away across the white mud-flat, level as a runway, a large saker has just thrown up into the air after a fast run in at a pigeon. With a few very fast wing flaps it has turned round, swung down again and put in short of some twigs of brush: the pigeon in refuge. Before they can say it the Landrover is rushing across the white flats. A tremendous flapping behind the bush and a man is out of the car, pulling off his head cloth and stooping as he runs in. The second car, seeing that all must be well, moves forward to join them.

A newly trapped saker is an ugly customer even though impeded with a foot caught up in the snare. The essential thing is to pick her up as quickly as possible before any feathers are broken. A cloth is thrown over her spread-eagled wings and the trapper closes them in his hands with a steady firm movement. He brings his thumbs together over the hawk's back, and his fingers round her chest, gripping the thighs between the ring and little fingers to stop the hawk clawing and damaging herself. If there is a hood to hand, somebody else will get it on her head at which, with the world blacked out, the hawk will forget her visual fears and get down to some energetic attempts to bite the fingers round her chest.

To secure the hawk and keep her still until she can be jessed and left hooded in peace on a block, she is put in a strait-jacket (*'aba*, cloak): a simple square of cloth, the two top corners of which are brought across and sewn down a central seam. The hawk's shoulders are pushed into the pockets thus created and the cloth wrapped around her, being tied tight with a belt which keeps her legs and feet flat against her tail, in the position to all intents and purposes of a dead bird laid down on a table.

Once back at the camp, there will be many felicitations and excited comparisons with other birds already trapped. Jesses are fitted, though makeshift cotton ones are often sewn on at this stage and the hawk is put in a quiet place on a low block made from a bag full of sand or gravel against which she can do herself no harm. Should she bate off, someone will gingerly try to get her up again. Keeping her feathers intact is of the utmost importance as a damaged hawk, even though the breaks are easily repaired, finds a disproportionately low price. If there is no hood available, the hawk will be seeled or, in plain language, its eyelids be sewn up. This apparently cruel procedure, which was standard practice in the West until the use of the hood was imported from the East in the Middle Ages, in fact causes the hawk no lasting discomfort and certainly saves her from much of the stress and trauma consequent on capture. A seeled hawk does not scratch or otherwise worry at her eyes and when the threads are removed the lids heal quickly.

To seel a hawk, a thread is passed through the upper edge of each lower eyelid – hawks, as opposed to humans, move the lower lid up to close their eyes. The threads are pulled through until the ends, which are knotted, are firm against the rim of the eyelid and they are then tied tightly together across the skull. This lifts up the eyelids and so the hawk cannot see. This is a very delicate operation and usually must be carried out in trying conditions. The hawk must be firmly held in an '*aba* by one person while another keeps a tight hold of her beak to keep her head still. The third operates, judging carefully where to put the thread. If he passes it through the lid too low, the skin may tear. The right place is just below the roots of the eyelashes. If done properly seeling has exactly the same sedative effect as the hood.

At the end of the day the trappers return from their last attempts to catch those birds which were looking for a meal before finding a roost for the night. They take a place in the circle of men drinking tea and coffee around the fire. One may tell how an eagle came up from nowhere and scared off a hawk which was just beginning to take an interest in his pigeon; another how his pigeon flew so well he lost it; how one sat all day watching a net trap and another how he caught a harrier and tried to use it as a decoy. The stories will be long and dramatic but inevitably as the newly trapped birds are discussed, talk will turn to a debate of the marks and qualities of a good hawk.

A fresh hawk is like a new slate and the good falconer sets down his training on it in a light and brief hand of his own. If the hawk is passed from one to another there is a danger of crossings out and such scars can be very difficult to erase. There is therefore no market for trained hawks and every hawk is bought untested and so something of a gamble. Hawks can be slow to show the symptoms of disease and an apparently fit bird can be dead within the week. A sick bird which took the pigeon because it was an easy meal in her fear and alarm until tamed, can easily look as tough and aggressive as her healthy neighbour. But while, in the last analysis, every falconer and trapper will admit that 'handsome is as handsome does', the Arabs have a ready eye for the signs which experience has shown are the hallmarks of a successful flier in the field. Secondarily, the professional trappers develop a sharp knowledge of the local likes and dislikes among their clients and, as they may see many hundreds of hawks, they quickly reach a shrewd understanding which a falconer in the West could take years to achieve.

To the Arabs, a hawk should have a small head but a full and arresting eye. Her neck should be long and drawn from the broad shoulders of a wide and powerful back; her legs be set wide apart and her feet have a long-toed span to give a large and strong grasp. In short, she should have the features indicative of strength and speed. Cosmetically, pale hawks are generally preferred to dark ones, reds taking precedence over browns. There is no real rhyme or reason for this, unless perhaps a rarity value attaching to lighter colours. So a trapper with a very dark saker will often make an undue performance over presenting it, declaiming that sheikh so-and-so had a particularly excellent black saker which killed all she was shown, and that, among the cognoscenti, dark ones are recognized as particularly promising. But this banter will cut no ice with the serious falconer who will be much more interested in learning how long the bird has been in captivity so that, by judging her weight and muscle, he can estimate her condition when trapped and, by the by, how many others have passed up the chance to buy her. Looking carefully at her eyes, if they are seeled, is a good indication of this. After a week or ten days, the eyelids reject the cotton thread and begin to open. Though this may look messy, the skin mostly heals as well as when the thread is cut and drawn out earlier.

In temperament the new hawk should show a wild-eyed self-confidence. Docility in an untrained hawk is often a sign that she lacks mettle; and uncontrollable wildness and aggression can be a symptom of waywardness. A bareheaded hawk when she bates from the fist, should bring herself up again quickly with a few wing beats, whereas the one which hangs by her jesses and fights like a tiger when lifted back by the falconer is a much harder nut to crack. If, however, the trainer successfully tames a hawk like this, she may well prove an aggressive and determined hunter. But, as in other walks of life, the bright, amenable pupil usually ends up as the apple of teacher's eye.

The intricacies of buying and selling hawks are bewildering and would almost be worth a book of their own. Because everyone wants a fresh hawk and they are trapped before their quarry is settled in, no performance data or track records are available. Everyone knows that all the comments and claims made about new hawks are shots in the dark, but considerations of 'face' prevent a buyer from trying to call the dealer's bluff. The latter makes a claim to knowledge, no matter how spurious, and the falconer is constrained to show he knows better. With businessmen also competing for hawks to use as presents, and all the oil wealth to be spent on them, it is a seller's market. The price of a hawk which has been with a dealer for a long time will rise to cover any other reason for her failure to find a buyer. Alternatively, a dealer may make a show of beginning to train some of his stock, saying that those birds are not for sale but for his own use. The more indifferent an owner is to an offer, the more he is supposed to value the hawk and she thus becomes irresistible to the buyer. Not only does he want the bird, but he wants to defeat the owner. There are loud scenes, reconciliations, appeals to extraneous and emotional motives, pleas for favours and for justice. The deal cannot be made until both parties are convinced that each has robbed the other.

A real difficulty arises when a villager by some fortune traps a serviceable but unexceptional hawk, a haggard peregrine tiercel for instance, and not understanding its real value, is convinced that he has the bird of the year and the answer to his financial problems. He frequents the places where hawks are sold and refuses to consider offers of less than twice the hawk's best price. The reverse may also happen. The red shahin which I mentioned above, which I bought for £3.50, I had expected would be far more expensive, perhaps £50–60. I must add that this price was fixed by a bedouin friend.

These days of trapping are energetic and happy. For anyone, the excitement and suspense of trapping a hawk is almost as great as that when she is first flown at quarry. It is a chance to see the wild bird, fit and untainted by human contact, give a demonstration of her natural excellence. The onlooker is torn between his anxiety for the outcome and his day dreams of the same hawk, trained and confident, leaving his fist for a similar chase. But for the Arab nowadays, all this is heightened by the enormous financial prize which rewards a good catch. After the enervating summer months, the migrations herald the return of better times for the bedouin and their animals and they bring promise of the hunting to which they are addicted. In these times when the pattern of nomadic life has so radically changed there are not many occasions when the men can go off from their families to recapture the atmosphere of the old days.

The camps also attract visitors: Syrian trappers and dealers from Damascus and Rahaibah, sheikhs with their retainers and falconers who have come to trap birds for their own use, emissaries from rulers or emirs further afield, come to buy. All these bring news and stories from all parts of the bedouin world, gossip about relatives working in Saudi Arabia or the Gulf. In the society of the desert where books and television are rare, talking is the staple pastime and as conversationalists the Arabs excel. An amazing peregrine trapped in Mesopotamia, a Qatari hawking expedition to Pakistan, the grazing in Kuwait, wild and fantastic stories of night life in distant capitals and, of course, politics, the politics of the tribe and of the Arab world as a whole: all these issues and many more will keep the trappers up deep into the night; the hawks as well, nervously alert behind their hoods as they are initiated into a world which till the following spring will be theirs also.

OPPOSITE PAGE *Hooded saker*

Training

'Obedience is the Mother of Safety and the Wife of Success.'
Aeschylus.

Before he can train his hawk, the falconer needs a few items of gear and other accessories which are known collectively in English hawking jargon as furniture. I describe them below seriatim for clarity, and explain the uses to which they are put more fully as they occur in the paragraphs on training.

Jesses The first necessity is to be able to hold the hawk securely on the hand and this is done by means of the jesses (*sbuq*), short thongs which are attached to each leg. Five or six inches in length (13–15 cm), made of plaited cotton, silk or nylon, they are never taken off the hawk except to be replaced by new ones. The jesses are held between fingers and thumb and when the hawk bates, or flies in fear from the fist, she hangs suspended for a second then, with a few flaps of her wings, regains her perch on the falconer's hand.

Leash The ends of the jesses are tied to the leash, or lead (*mursil*), which has three sections, two of about nine inches (23 cm) and the third of about two feet (60 cm). The first short length joins the jesses and is in turn attached to a brass swivel (*midwar*) which joins it to the long section but allows the hawk to turn and move about without the leash becoming twisted. The other short section is kept permanently tied to the block.

BLOCK (*wakr*) where the hawk spends most of her time. This resembles an enormous golf tee or drawing pin. Its round top is padded with carpet or leather and the wooden stem is decorated by turning on a lathe and usually with brass nails as well. Old blocks are sometimes beautifully finished with silver wire inlay. The stem ends in a steel spike of about a foot (30 cm) so that when the block is stuck in the ground, depending on the hardness of the surface, the hawk stands at a height of approximately one and half feet (46 cm). The flat top enables the hawk to spread its weight across the span of the foot, for not being tree-perching birds they find having their weight concentrated on the ball of the foot very uncomfortable, if for any length of time.

THE HOOD (*burqa'*), is an important aid to the Arab falconer who has no darkened and secluded mews where he can leave his hawk when he has no use for her, in the earlier stages of training. Like most other diurnal birds, a hawk will sit quietly enough if left alone in a dark place with no window or other source of light at which to fly. If in pitch-black darkness, she will sit almost motionless with no thought to fly for fear of breaking feathers. This is the situation which the hood neatly creates by blindfolding the hawk. It follows then that a hood to be effective must be a perfect fit, admitting no light at the opening through which the beak passes and yet standing well clear of the eyes and ears to avoid irritation. Once on, the hood is closed to a tight fit by pulling the longer of the two sets of 'braces' which are fitted at the back. Pulling the shorter pair, conversely, opens it again so that it may be pulled off by the tassel or thong which is attached to the hood's crown. The efficacy of the hood is such that an unruly or excited hawk can be calmed and pacified in seconds. This is important not only in training, but also in the field where the flapping of a hawk's wing could alert unspotted game which then lies low and is missed. Contrary to expectation, the effect of the hood on the hawk's eyes appears to be nil. A hawk which has been hooded for hours can be unhooded in the strongest desert sunlight and react instantly to the distant form of a houbara or hare making hard for the horizon. Keeping a nervous hawk hooded is kinder than letting her constantly bate and exhaust herself with fear.

THE LURE (*tilwah*) is the falconer's semaphore. It is a bundle of houbara wings and the falconer swings it in vertical circles on a line to signal to the hawk that she should return. If the hawk misses her quarry, she can see the lure being swung – its black and white feathers make it easily visible – and she flies to it in the expectation of the meal she has just missed.

THE GLOVE (*dass* or *kaff*) The last item necessary is some protection for the left hand on which the hawk is carried. In northern Arabia a stout glove of three or five fingers is used but in the lower Gulf, a cuff or *manqalah*. This is simply a cylinder of carpet or canvas, much like a woman's muff, which is worn on the left arm. A loop is sewn onto one end to take the leash and the falconer feeds the hawk by holding the meat in his fingers which protrude at the other and open end.

HAWKING BAG (*mikhla*) European hawking bags can be elaborate, full of little pockets, tucks and tassels. But the nomad, true to his taste for the spare and functional, uses a plain square sack. It is made of canvas or cotton and, worn tight into the armpit on a short string, it reaches down to the waist. The *mikhla* carries the lure, a knife, a spare hood and leash and, with luck, the kill. The root of the name means 'hiding': the falconer sometimes spreads the bag over the hawk's feet and the kill beneath to convince her that it has disappeared. She can then be picked up without fuss or complaint.

In many parts of the world, falconers tie bells to their hawks' legs to help locate them in cover, or if they are lost. The bedouin, however, do not use them, reasonably enough as there is little cover in the desert and, if you can hear the bell, you can probably see the hawk anyway.

TRAINING The key to the management and flying of hawks lies in condition. They cannot be punished or starved into compliance, nor do they ever behave or perform in a way which may be taken as intended to please. Instead, hawks must be brought to a physical and mental condition in which they may, with luck, be counted on to react predictably in a carefully controlled situation where the options open to the hawk are heavily weighted in favour of the course desired by the falconer. That is the principle of preparing hawks for the field. Training is the usual word for this process, but apart from overcoming her natural fear of man and accepting food from him, the hawk is required to do little which is not but a small adaptation or extension of her natural instinct. The falconer needs to be able to take his hawk to the field without upset, to show her quarry and to be able to rely on her to return to him when called, should she fail to kill. The hawk will not retrieve in the event of a kill and indeed she would not be capable of carrying the quarry at which the Arab falconer flies her. The ideal hawk is the newly trapped bird taken from the snare or net by the falconer who will train and fly her. If such a bird develops faults as a result of mishandling, the falconer has none but himself to blame. Few, however, can be expected to manage a hawk without any mishap or mistake that may alienate or offend her, so the fewer hands through which she passes before the falconer's the better. But in practice, she must first be dealt with by the excited trapper whose one aim is to keep her from breaking any feathers before getting a good price for her; then by any number of the men at the camp who will help hold and feed her before she is passed on to the trader who takes her off to the markets in Damascus, or delivers her to a buyer in the Gulf or Saudi Arabia. Finally, having been handled by possible buyers or admiring friends at the owner's *majlis*, she may be passed to the retainer who is to train and look after her. The hawk will have been kept hooded or seeled and the handling she has suffered will have gone a long way to break down the first barriers of fear. She will be used to being hooded and will feed without much hesitation.

As this will have brought her through the early stages of training and manning (the English term for taming a hawk), let us return for a moment to that rare and much to be wished for coincidence of the falconer trapping his own bird.

The hawk, head drooped under the unaccustomed weight of the hood, stands on her block with wings hanging loosely by her sides. At the least near sound she turns and hisses, perhaps, and especially if she is a peregrine, bating blindly from her perch. She is left as quiet as possible (which is not very, in a tent) for twelve or twenty-four hours to sharpen her appetite and then is lifted, still hooded, onto the fist. At this strange contact she will almost certainly bate, claw and wrench the fist and, finding herself anchored by the jesses, bite furiously at these and at the glove. The falconer's one aim is to get her to feed and, having some light and easily-pulled meat, will try to see that when she bites she closes on this. She may swallow it and continue, but being contrary and having no idea of what is meant, she will probably flick it away irritably. The falconer teases her toes with the meat and perseveres.

At this stage, it is a moot point whether she would be more likely to feed if she could but see the food, or whether unhooding her would unleash a new bout of bating and fretting. To this, of course, there is no answer and either may be the consequence of taking the hood off, though with the more nervous peregrines, unhooding is usually disastrous. At length, whether by ruse, persuasion, unhooding or teasing her, the hawk will be brought to feed and be allowed a small meal: a crop the size of a pullet's egg.

Outside Arabia this first lesson is best given alone in a room where, particularly if the light is subdued, there are no distractions to increase the hawk's alarm. But for the bedouin still living in a tent, such seclusion is not to be had and there may be any number of people, from a few children of the household to twenty or thirty men in a sheikh's *majlis*, under whose critical eyes the falconer must reach the same successful conclusion to the session.

Once a hawk has been induced to accept a meal on the fist, she has made the initial and greatest compromise from which the rest of her training follows without much ado. She has decided that in the balance of fear and food, because the falconer has been so gentle, the latter outweighs her natural disposition to escape. This routine of dilemma, calculation and concession is reproduced day after day until the hawk will feed quite contentedly in the midst of a scene which previously would have caused her acute anxiety.

It is a progression of trauma which inures her to her new and alien world. If she is pushed too far too fast, her great reserves of stubbornness will exasperate her falconer, but with tact and careful use of the hood to forestall a crisis he will overcome her fears. With time, she will begin to sit quietly without the preoccupation of food. To the desert falconer tameness in his hawks is of paramount importance: so crowded and uneven is his life that unless they are well manned, they quickly deteriorate. Physically, their feathers become scuffed or broken through endless bating from the block and this seriously affects their efficiency in the field. Psychologically they can become suspicious and resentful. A new hawk, once her initial fear is overcome, is a gentle character, but if she is mishandled she can quickly develop the vices of snatching and biting. She views people not so much with fear as with distrust and dislike and such a hawk is nothing but a burden. The Arab, then, makes his new hawk his constant companion to hasten the process of manning. There is honour in being seen with a hawk, especially today when they are so expensive; but the underlying reason for the hawk's frequent appearance in the bazaar or *majlis* is the continuing need to keep her manned.

There is even some social disgrace in having an unruly hawk. It shows that you have failed in your undertaking and, more generally, that you have not been what most bedouin would like to consider themselves: a good falconer. I was very embarrassed when a sheikh with whom I was staying, wheeled round from his place by the fire and curtly asked one of the slaves what a wild hawk was doing in his tent. I felt even worse when he had to let the matter pass. At the far end of the tent one of the hawks had bated twice. It was his guest's and mine.

Any group of European falconers will argue all night about the advantages and disadvantages of touching or stroking a hawk to increase her tameness. On the one hand hawks certainly do not enjoy being touched, as, say, a cat may, and there is some risk of the weatherproof bloom being taken off their feathers. Further, some falconers argue that they do not like their birds to be too tame as this does not appear to affect their success in the field, and can put them at risk from trappers or shooting people if lost. On the other side there is a school of thought which reasons that a bird which is not afraid of being touched is easier to manage and is, in general, a greater source of pleasure and pride to the falconer. Considerations of weather and rain are of course of less account in the desert than in Europe, and indeed the saker, a desert hawk, oils its feathers much less than the peregrine. But, in any case, the Arab falconer sides firmly with the latter point of view and his hawks are relentlessly stroked with firm downward sweeps of the palm from the head to the thigh, following the contours of the wings and chest. Sakers respond very well to this treatment: rather than being simply chained to the falconer's fist and left to grow restless and anxious, the bird's interest and attention is scarcely allowed to wander. She is preoccupied with the movement of the falconer's hand and the food the falconer produces from his bag.

The Arabs also differ from falconers in other parts of the world by teaching their hawks to know their names. Indian and European falconers do name their hawks but do not train them to answer to them. The Japanese traditionally used to believe that hawks are afraid of the human voice and called them by clapping the lid of the box from which their hawks were fed. A bedouin, however, from the earliest stages of training shouts out his hawk's name as she feeds, even when she is sitting hooded on his fist. The hawks seem to accept these unnerving attentions philosophically as part of the total strangeness which surrounds them. In time they become conditioned and even when hooded will bate from their blocks when called by name. I used to reflect when sitting with bedouin in a tent that it was a measure of my foreigner's isolation that, even though they had seen me train my own hawks, had I bellowed out any of their names, someone would have dropped his coffee cup at the shock of the noise. But the tribesman's piercing shouts rend the even hum of conversation without anyone turning a hair. The bedouin have been doing this since long before Pavlov and it is interesting that they do not bother to speak the hawk's name but always shout it. It is a shout, after all, which she will hear when her falconer is calling her in earnest. The names they give, like hounds' names, are short and easy to call. *Sabah* (Morning), *Garrash* (Grinder), *Dhib* (Wolf) and *Sultan* are but a few.

As soon as the hawk is tame enough to sit still for a few minutes without being fed, the falconer begins to coax her to step onto his fist for her food from the ground or the block. At first, a bird which has proved her mastery of flight by staying alive in the wild will try every possible means of getting the proffered food without actually flying to it – as if she had forgotten the use of her wings. She will lean out so far that she almost falls and has to flap quickly to stop herself from overbalancing. She will turn her head from one side to the other to see if the task looks any easier from another angle. In frustration she resigns herself to the 'displacement activity' of pacing up and down on her block or pulling at her jesses with her beak. Finally, however, she does make an inspired leap and lands on the fist. Once this mental block has been removed, progress will be more rapid and in a matter of days she will be jumping the full length of her leash, as soon as the fist is held up.

When she will come readily, it is time to introduce the hawk to the lure. The falconer takes a pair or two pairs of dried houbara wings and sews them tightly together. Then he ties to this lure a freshly killed dove and throws it down beside her. She will jump onto this and begin to feed and the falconer, after helping her with a few mouthfuls, lifts her onto the glove or *manqalah* to finish her meal there. She will quickly come to realize the significance of the lure and will fly to it from increasing distances. Each time she is rewarded with a meal. Her readiness to come to the lure is more a reflection of her increasing tameness than of any hunger. Certainly, she will not come if she has been overfed but neither will she because she is starving. The falconer aims to make these 'calling off' exercises of coming to the lure an easy progression in which the hawk finds each stage of the increasing distances she must fly natural and unhurried. For calling lessons outside, a long cotton or nylon line is tied to the jesses or leash.

If, on being taken outside, the hawk seems paralysed with the same kind of helplessness which plagued her first jumping lessons, or if when coming a distance of about twenty yards, she persistently veers off (probably in fright at seeing the falconer's form filling her field of vision fast as she flies up to him), the Arab falconer may substitute a live pigeon for the lure, a trick which will bring her sure enough, if she is the least bit keen and ready to fly.

When she will come fifty yards or so to the swung lure, she may be flown free, though usually over a shorter distance for the first few times, unless the falconer has nerves of steel. In some areas, particularly in the Lower Gulf, hawks are trapped and trained well before the arrival of the houbara, and so these daily calling off exercises may continue for a number of weeks, in the course of which the hawks can reach a high standard of obedience.

The Arabs do not make a regular practice of 'stooping their hawks to the lure'. This is a way of exercising a hawk but not a means of improving her skills as a hunter. The falconer jerks the lure out of the hawk's path just before she strikes it. Then he continues to swing it and the hawk banks to make another attack. This may be repeated a good many times to keep her on the wing and flying hard. Some Arab falconers do make their hawks stoop once or twice, but they say that to make her do more than this develops in her a distrust of the lure. This is not the experience of those in other countries who regularly stoop their hawks to make them fit. But the desert tradition is strong and the Arabs are convinced of its worth.

The hawk's training, however, cannot be considered done until she has been introduced to quarry and successfully dealt with it, until, in the English falconer's terms, she is entered and made. A manned and obedient hawk can be shown a hare or houbara and take not the least notice, so conditioned has she become to the inevitability of being fed on the fist or the lure. Some, of course, go from the first and kill in fine style, as though their period of training has been nothing but an unwelcome interruption.

The entering of passage hawks to quarry is an easier business than that of eyasses which have never killed in their lives. But the stage management must nonetheless be carefully handled when a houbara is involved, if the hawk is not to come off the worse and be alienated from so large an adversary in the future. This means bagged quarry. One of the previous season's hawks is used to catch, but not kill, a houbara; it may be winged with a shotgun or, if a hare is used, it is taken with a saluki. If too much bounce is left in the victim, the Arab takes a few primary feathers out of one wing and then releases it in full view of the hawk in an open space where it cannot make easy cover. Few hawks can resist such a chance and the unfortunate creature's ordeal is quickly over. A repeat performance with another, but at great range, is probably all that is needed to wed the hawk to the quarry. Thereafter, the black-and-white flash of the houbara's wing beat will be enough to send her rocketing from the fist in the expectation of the full crop of warm meat which she got on the previous occasion. Indeed, until she is absolutely reliable, it is unsafe to deprive her of a full meal from her kill lest she consider the game profitless and lose heart. A hawk entered in this way will take on houbara at a range of up to half a mile.

Mention, perhaps, may be conveniently made here of the Arab practice of flying hawks at pigeons thrown out by the falconer. This short-cut to sport is a sure guarantee of exercise for the hawk and great excitement for the onlookers, though often resulting in the loss of the hawk. Contrary to common supposition, the pigeon, if flown wild, is anything but a sensible quarry for the trained hawk. It is not found in country open enough to provide scope for the pursuer and almost invariably flies hard and straight – a tactic most likely to cause the hawk to turn tail or run out of breath once she is far, far from the falconer and his lure. But a house pigeon thrown out from a standing start under a hawk on the wing is a matter a little more evenly balanced and, as all hawks love eating pigeons which in the wild they catch by stealth and sudden attack, a breathtaking pursuit follows. But this is not hunting and hawks thus pampered may refuse real quarry or, worse, the lure.

OPPOSITE PAGE *Furniture: block, manqalah, hood, leash and jesses*

QUARRY

'I am escaped with the skin of my teeth.' Job.

HOUBARA

Macqueen's bustard (*Chlamydotis undulata macqueenii*)

THE HOUBARA IS A SMALLER MEMBER OF THE BUSTARD FAMILY. MALES weigh up to five and a half pounds (2·5 kg) in comparison with the great bustard, of which the males go up to twenty-four pounds (10·8 kg). Apart from in size, it is distinguished from the rest by the decoration of long and fine black and white feathers down its neck. Both the great bustard (*Otis tarda tarda*), and the Arabian bustard (*Ardeotis arabs arabs*) are found in Arabia. The former was once hunted with greyhounds on Salisbury Plain in England, but became extinct in that country by about 1835. It has southern limits to its distribution on the open plains of North Arabia, south to the latitudes of Damascus and Baghdad. The Arabian bustard is found as far down as the Yemen where it is known as *lawwam*; but both races are now very rare. The desert Arabs call any bustard larger than the houbara, *hubrum*. I have heard reports of *hubrum* on the shores of the Gulf and in Iran, but as the word does not distinguish between the Arabian and Great, nor yet the Indian allied form *nigriceps*, it is impossible without seeing a carcase to be more definite. Wild stories are occasionally ventured about sakers taking on a *hubrum* in a fit of bravado, but the half-closed eyes of the other listeners perhaps betray such tales as fantasy and greed.

The more manageable houbara, at a fraction of the size, makes for more realistic quarry, though nonetheless it still presents some tricky problems for the hawk which only weighs half as much or less. The houbara is found throughout North Arabia, the areas of Qasim and Najd in the centre and down the eastern seaboard from Kuwait to Muscat. Singly, or in small family groups of four or five, they frequent any undisturbed and open place where they can find the artemisia shoots and grubs of which they are so fond. They feed also on beetles, locusts and other insects and will take small lizards.

As with the recipe given for the cooking of hares, the first problem with hawking the houbara is to find it. Very shy and retiring, they will run using the cover of wadi beds and dead ground way ahead of the ranging hunters and, if surprised at close range, lie flat with their necks stretched out along the ground in the hope of not being spotted. In this, they are relying on the fact that hawks can be very slow to notice unmoving objects. Once the houbara takes to its wings, however, its size and the flashes of white on its slow-beating wings make it an easy target for the hawk to spot at enormous distances.

But in spite of its unwieldy appearance, the houbara is no mean flyer and given a good start will take a hawk a considerable distance. At closer range, it will often prefer to see things out on the ground, and here its size and determined defence will shake off any but the most resolute attack. In extremes, the houbara will sometimes defend itself by squirting out slimy faeces to a range of two or three feet which, if the hawk is hit, will render her helpless with gummed up and clammy feathers on which she cannot fly. Hit in the eyes, she is temporarily blinded and the houbara, assuming the hawk has not already got a hold of it, can make good its escape – the last resort in self-defence. The Arabs call this mess of the houbara's, *tamal.* It can be removed by careful washing.

Though the houbara does breed in Arabia, and I have myself found eggs in early March, south of the Buraimi Oasis in Oman, the majority are winter visitors from the north. They arrive late in November or early December when the desert is beginning to come alive after the first rains, and are mostly gone by middle or late March.

A disappointment for the foreigner is that the houbara, looking at a distance about the size of a small hen turkey, is a very lean and spare creature when plucked. But to the bedouin this is no surprise and the meat when boiled and put on rice is regarded by them as second to none. The houbara holds a singular fascination for the tribesmen. Its cunning and courage in defending itself are the subjects of endless stories. They also believe, evidence of its remarkable nature, that it is the only bird which has twenty-four feathers in its tail.

KAIROWAN

Stone curlew (*Burhinus oedicnemus saharae*)

There is much in the stone curlew to suggest a junior branch of the bustard family, though of course in scientific and ornithological terms this is not the case. Like the houbara, the stone curlew has no hind toe and so makes a similar if more spindly track. It has much the same diet and so frequents the same habitat. It shares the larger bird's yellow bright-button eyes and also shows white wing flashes in flight. But it is very much smaller: Meinertzhagen in his *Birds of Arabia* gives the following wing measurements: 235mm for females and 252mm for males comparing with the houbara's 345–430 mm. Such is their ability as trackers that they are usually able to do this without much trouble.

Like the houbara, the stone curlew is a timid unobtrusive figure, lying up in the sanctuary of some cluster of bushes during the greater part of the day, only emerging in the dawn, twilight or night-time. The bedouin reckon that it is not a great traveller and if they find tracks not more than a day or so old, they quarter up an area about a kilometre square and expect to find the bird which made them.

But in flight, its agility and bat-like jinks can cause a large hawk quite some exercise in twisting and turning after it. For this reason the tiercel is the ideal antagonist to play against it – his shorter wings and lightness make him quicker on the turn. Once struck, resistance is over: an easy kill if the slip has been short and the hawk sharp.

ARNAB

Hare (*Lepus arabicus, Lepus capensis*)

The European hare (*Lepus europaeus*) is found in the Middle East but not at all frequently in the desert. There is some doubt as to whether or not its slightly smaller congener, *capensis*, is in fact a different species, or simply a race bordering its neighbour and blending with it in shared and merging geographical distribution. The true desert or Arabian hare is the smallest of the three and of a lighter colour more suited to its surroundings. Some say that it is faster than the European, but I very much doubt this. Similar claims are made for the desert's other athletes, the Arab horse and the saluki. But size in nature is usually a good indicator of speed and the Arab horse competing with a thoroughbred, or a saluki with a greyhound, both larger animals, are quickly left in no doubt as to speed. What may indeed be true of the hare, as it is with the horse and saluki, is that it has superior endurance, born of a harder and more open habitat. But I know of no experiments done to establish this in the hare's case.

Speed aside, which is nonetheless considerable with all the familiar tricks of turning through ninety degrees at a full gallop, the desert hare is remarkable for its ears – they are huge! The role of these cavalier ears appears to be to regulate the animal's body temperature. Schmidt-Nielson, an expert on the physiology of desert animals, argues that the cloudless desert sky has a high capacity for receiving radiated heat and may thus be considered 'cold'. A hare crouched in the shade, and preferably in a hollow to escape the worst of the surface heat from the earth, is protected from the direct heat of the sun. In these conditions the ears come into play, their large surface area acts as an efficient radiator of heat to the apparently 'colder' sky. Hares do certainly spend most of the day in a shallow and well-shaded form. On occasions, they make use of other animals' burrows or holes, but do not make a regular habit of digging for themselves.

The hare, flushed from its form, makes an exciting and taxing flight for a hawk as its darts and jinks from cover to cover. Though only the size of the European rabbit, its true hare shape gives it powerful back legs with which to kick out at the attacker and for this reason many peregrines who loathe affrays on the ground are shy of flying this quarry.

IDMI

Arabian gazelle (*Gazella gazella arabica*)

The gazelle is an exquisite, liquid-eyed toy of an antelope standing only about two feet six inches (67cm) at the shoulder. Its relative, *Gazella dorcas* (*afri*), is also found in Arabia, but is slightly smaller. While in the past large herds were not uncommon and a hunting expedition could scarcely fail to see tens if not scores, during the last thirty years they have fallen a sad casualty to the catastrophe of the automatic gun and the jeep. To see a gazelle in any part of Arabia now makes something of a red-letter day. This animal is one of many desert mammals classified as 'endangered'. Conservation practices recommended for this species include surveys and research to determine populations, forage preferences and habitat requirements. These may be all very well in theory but one wonders just how practical they are when one considers the elusive nature of the subject, and the immense range covered by this fleet-footed gazelle.

Its status as a quarry for the hawk is at best problematic and I only include it because it is so often mentioned in other accounts of Arabia. In fact, among many hundreds of falconers and tribesmen across Arabia, I have never met one who has said that he himself has taken a gazelle with a hawk alone. Nonetheless, gazelle hawking is frequently described in references to Arab falconry and one can meet any number of people who will explain how it should be done. Needless to say, it is a ruthless business requiring some fearful entering. The gazelle, of course, can hardly form part of the wild hawk's diet. How would she kill it? Its head is too large for a hawk's foot to grasp and is protected by a hard skull. Should she bind to any other part of its body, she would surely be shaken off or badly injured by the gazelle's rolling on her. It is possible that the occasional calf may fall prey to a hawk, though gazelle are conscientious mothers in the defence of their young, and such a case would be listed under 'miscellaneous' in the hawk's game book. So if the hawk does not herself think of the gazelle as a food parcel, she must be taught to make this discovery. A small gazelle is caught and the hawk trained to feed from strips of meat tied to its head and horns. After several of these sessions the hawk will fly to a tethered gazelle expecting to find food on its head: a process much the same as teaching her to come to the lure.

In the field, the falconer, taking with him a saluki, stalks up a single or small herd of gazelle and then slips the hawk, usually a strong female saker. She quickly closes with the gazelle and to look for food on its head, makes several attempts to bind to it, putting in a succession of short stoops from all sides. The gazelle, jinking to avoid these attacks, will be forced off its direct route of escape and the saluki will then have a better chance to close with it. I do not myself believe accounts of sakers actually striking the gazelle's head with her stoops, time and again, until the unfortunate creature, confused and concussed, sinks to the ground exhausted. A more likely conclusion to the flight would be the saluki bringing it down by the legs or neck and the hawk then landing alongside; mercifully, this practice is now ended by the scarcity of quarry.

There are many other species of bird found in the desert during the hawking season which could give trained hawks exciting flights. But they are protected by the inhibiting force of the bedouin's traditions and the limitations of his techniques. A few partridge, waders and sandgrouse are taken but are usually considered too fast for hawks flown from the fist, or not worth the trouble. Lawrence al Sha'alan once told me how he had a small sakeret which he intended to release as he thought he was too small to take usual quarry. Driving in the desert he came across a flock of sandgrouse and slipped the sakeret simply to see him fly well on his way. To his amazement, the sakeret killed one in fine style. Lawrence picked him up again, but try as he might, he could never get him to catch another. The incident was put down to chance and his view that sandgrouse are an impossible quarry took firm anchor again.

I saw a sandgrouse killed with a female saker in Oman but this was considered more a feat of the falconer than of the hawk. We had seen a flock put in a little way ahead and drove over thinking to have a shot at one. Getting out of the car, they got up before the gun was ready. We lit cigarettes and began to talk of returning home, when suddenly at our feet a sandgrouse exploded into flight. One of the bedouin unhooded and slipped his hawk before the sandgrouse was over the roof of the Landrover just behind us. We raced round the car and found the hawk on her kill not twenty yards away.

THE FIELD

'Doth the hawk fly by thy wisdom?' Job.

IN LATE NOVEMBER, WITH THE ONSET OF COLD WEATHER, THE HAWKING season begins, in fits and starts at first as reports come in from shepherds, camelmen and travellers of tracks or quarry seen in the desert: new arrivals from the north. By this time the hawks will have been in training for a month or more and, in recent weeks, flown daily during calling-off exercises to the lure. The first news of quarry brings great excitement and a party of falconers will rush off to open their accounts with the season's game.

These short early outings, which last only an afternoon or the inside of a day, give the falconers a chance to get some impression of their birds' abilities, and to bring on or enter those who are slow starters. Nowadays, the Arabs go hawking in jeeps, pickups or Landrovers, but in the past they had to make do with camels or, for short distances, horses. There was some advantage in the old style. Quarry were accustomed to camels browsing in the desert and would not take fright at their appearance. So the falconers could approach them much more easily than is now the case when they travel by car. When riding, if the hawk was not carried on the left hand, she was put on a pad behind the saddle where, being hooded, she would sit quite still. I once asked a bedouin who had no car, how he transported his family of very active cats when on the move. His answer was brief but complete: he held up both hands, flexing the gnarled fingers like gripping claws, and roared with laughter. The same clearly applied to hawks. They still face similar problems of balance when carried in the backs of open cars. But they seem to learn quickly that they must at all costs keep their wings in tight to their bodies and not use them to steady themselves as, if they did, they would be blown off their feet. I have seen them crouched behind spare tyres and bedding, riding out the lurching and bumping of the car, as still as decoys. Afternoon trips in response to a definite sighting not far from home require no more gear than the falconer, his hawk and his hawking bag. A shotgun or rifle, like the bedouin's shadow, goes too.

If the sighting has been sure and the news brought in without delay, it will not be long before the falconers find the quarry. A group of three or four stone curlew would be a likely and useful start to the season. Once in the vicinity, the driver quarters up the watercourses and stretches of gravel or hard earth in-between; and all with him keep a careful eye open, not only for flushed birds ahead, but also just beside the car for tracks or the white splashes of droppings.

From careful observation, the bedouin have developed a close understanding of tracks and can interpret them with what to a stranger seems almost prophetic sureness. A short single file of tracks indicating a run, ending with two footprints together and a dropping, is a certain sign of a stone curlew's taking wing. In the case of the larger houbara, though the droppings may not be there, this can often be confirmed by the faint brush-marks on either side of the footmarks where its wings have scraped the ground when taking off. I have often been staggered at the Arabs' ability to spot these tracks from a car moving at as much as twenty miles an hour.

We were once hawking on the northern borders of Saudi Arabia, near Jabal 'Anaizah. The car shuddered to a halt with the driver standing on the foot brake, all jumped out and a heated discussion followed, held with faces bent over the ground. 'Do you not see them?' 'Well, no – what, where?' and then, pointed out with a finger, the faintest traces of footprints. A nail mark in the dust where the houbara's foot had been on pebbles, a clearer impression of one step a few inches further on and then, to my eyes, the trail was lost again in the gravel.

The clue that a stone curlew has taken wing need not dismay, as the bird usually lands again not many yards ahead. The tracker will look up and trying to read the bird's mind, work out where it would have headed for. If no obvious bush or patch of new grass gives him a hint of where the trail continues, he returns to quartering up the ground. Once flushed, there is no mistake. After the larks and pippits which have got up in front of the car, the stone curlew looks disproportionately large with its white and black-flashed

OPPOSITE PAGE *The lure*

primaries striking staccato beats as it dances higher into the air. In flight the curlew will sometimes shout out cries of alarm and this, with its long legs hanging down until it is properly airborne, makes it a somewhat ghoulish sight.

The next few minutes will pass in a flashing confusion which may leave a new guest wondering what to make, if anything, of the sport. It all seems so quick and over so suddenly. But in slow motion, the falconer whose turn it is unhoods his hawk (he will have had the braces open and ready) and pushes her clear of the car, round the door or through the window, simply on a raised arm if he is riding on the open back. She may bate instantly at the quarry and be off, or look about with a bobbing head bringing her gaze back to the prey. If she is uncertain, she may 'rouse' or shake her feathers, and mute, but with luck she will eventually go, dropping a few inches from the fist as she launches, then picking up on this momentum with strong stirring beats. At this point, the stone curlew must decide whether to seek safety in flight, or in the cover of the lee-side of a bush or bank. The strongest of the group will probably choose the first course and be away in no time, separating quickly to force a dilemma on the pursuer. The hawk makes on them quickly and as she does so, the tail marker, a young and inexperienced bird, or perhaps an elderly and infirm one, may lose its confidence and drop down like a plummet, chattering in alarm.

A stoop from the hawk at its place of refuge and the curlew jumps clear just as she passes at the bottom of her swing and is least able to change course. Off again; but the hawk has banked, turned and is after it. It puts in and the hawk dives, misses and standing on her tail, pumps hard with her wings to gain height and inhibit the curlew from taking off. She hovers momentarily to position herself and stoops. The quarry, it must be remembered, is superbly fit by comparison with the hawk whose only exercise has been a few dashes across an open space to a swung lure. I have seen curlew bounce and leap off the ground to escape repeated stoops, until the hawk, in short-breathed frustration, has landed or swung away to be lured down by the falconer. Should this happen, another hawk will be slipped, perhaps a peregrine. If the curlew is still grounded and has not taken the chance (of the first hawk's break in concentration) to bolt, the peregrine will home in with her rapid, shallow wing beats leaving no room for question about her superior speed. The stone curlew dithering behind a bush makes an easy target for her and she sweeps in, turns the corner of the bush on a sixpence and binds to her prey as it runs to take off.

If the whole flock stays in the air when flushed, a stern chase ensues; though the hawk's greater speed, if her breath lasts, soon tells. Sakers do not often make a point of seeking the advantage of height as a peregrine is inclined to. Flying down the quarry, they will bind to it and bring it flapping onto the ground. Once landed, the powerful toothed beak of the falcon is quickly brought into play: she leans down and with a pull dislocates the neck of the loser. Then, after regaining her breath, she begins to pluck the prey and feed from its neck and head.

The falconer 'makes in', approaches the hawk cautiously, stooping down to avoid standing over and frightening her. The hawk flies with only her jesses, so he takes hold of these and reattaches them to the leash to secure her, and then helps her feed with his knife. As he does this, he covers the greater part of the dead quarry with his hawking bag or by scooping sand over it. By cutting off the head, or producing another piece of meat for the hawk, he can get her to concentrate on this smaller bit and step onto his glove to eat it, forsaking the kill. If the falconer hopes to fly the hawk again that day, he will allow her only a few mouthfuls and some warm blood. Otherwise, the edge of her appetite may be blunted and she will lose her zest. Returning the hawk and bag to the car with much triumph, the party wheels on in search of the rest of the flock.

It is hard to communicate the impact of the explosion of excitement which occurs when quarry is sighted. The hours of tracking and cruising about the desert have their moments of climax and optimism, but are as much coloured by a grim determination and tooth-sucking silence. A houbara's wing on the skyline, or a hare bolting from the rumble of the car, ignites a fast burning fuse of activity. The car drops a gear and surges forward, hoods are loosened, men struggle to open a window or load a shotgun, instructions, counter instructions and with all the rocking and bumping over the rough ground – a suffusion of adrenalin which is sure to drown any attempts to photograph the occasion.

In the past, when true nomadism was the norm for most if not all bedouin, probably much more hawking was done in this literally off the cuff way. Hawks would have been ready during the long rides from camping site to camping site, and many a houbara or hare must have been picked up in the bow wave of the advancing tribe. Today, the Arab must take out his car and hawk with the intention of hunting. Then, being on the scene with a sharp set falcon was coincidental with the business of life for many days of the month. The same would have applied to the salukis who used to trot along in the moving shadow of the dromedary or mare, loose and ready to course any hare put up. The loss of this involuntary usefulness may be a reason why the population of pure blood and experienced salukis in Arabia is certainly declining. Not so many are ready to take a saluki out simply to course hares (and not shoot them) and, of course, there are not so many hares either.

But even in those days of abundant quarry, when hawks were worked far harder on more days of the season than they are now, special expeditions were mounted when the men could give themselves entirely to their hunting. The tradition is continued today and indeed for most Arab falconers such trips make the body and not merely the climax of the season.

The difficulty of finding quarry makes this especially true for those living on the shores of the Gulf. They must travel to find game. Like the trapping expeditions, these hawking parties can become important social functions – all the more so for the status of those who lead them. Sheikhs and princes, ministers, presidents and, in the case of Saudi Arabia, the king, all make elaborate hunting forays – from Qatar to Syria or Oman, from Dubai to Iran, Kuwait to Pakistan, Abu Dhabi to Iraq, from Saudi Arabia to Afghanistan or Morocco. VIP lounges at international airports throng with teams of retainers, friends, fellow tribesmen and cabinet ministers, all helping to get the hawks aboard the private VC 10 or Boeing. Lavish tents and kitchens are set up in remote valleys and beside them radio stations to keep the falconers in touch with home and international news. If events cannot be dealt with by telegram, the jet is sent to bring a messenger or an ambassador with an up to the minute briefing.

In 1975, I was disappointed not to be able to take up an invitation from Lawrence al Sha'alan to join the annual hawking holiday which he used to make with his brother-in-law, 'Abd Allah ibn 'Abd al-'Aziz, Commander of the National Guard in Saudi Arabia and brother of the king. But later I heard in graphic detail from the bedouin what I had missed. The party had comprised about eighty vehicles including two tank transporters carrying air-conditioned bungalows, a communications unit, electricity generators, mobile kitchens and a complete staff to organize the logistics and commissariat necessary to launch nearly a hundred hawks.

On large expeditions, the falconers fan out from the base camp in small parties of one or two cars and work the sectors allocated to them by the sheikh. On a level place well-broken with bushes they may leave the car and walk. With no setters or spaniels, this is the only way to put up quarry which is intent on lying low – a houbara seen ducking into a large shrubbery in the middle distance, or a hare which has bolted from one water course across the intervening gravel and safely into the cover of the next. For the rest, the party will stay put in the cars as with quarry so scarce there will be much ground to cover before finding it.

In the evening all return home to the camp and its fire where tea, coffee and more tea are being got ready. The hawks are well fed and set out on their blocks, and the men make a great ring in the firelight. Where there is a large party, visitors drift in like moths, attracted by the chance of conversation, news and perhaps a meal. They come to pay their respects to the sheikh, to present a petition for help or to ask for a decision in some local dispute.

I remember a vivid occasion on such an evening when an exile from our party's home country appeared from the darkness and joined the circle. Only a degree of restraint in the hitherto easy atmosphere alerted me to the probability that the guest was an out of the ordinary run of visitor. The reflex hospitality of the Arab and the vice-like social disciplines of his upbringing clenched tighter. The tension was dramatically understated by the leader of our group who was sitting next to me and spoke to me in English, which he knew the newcomer did not understand. He explained that the guest was a senior member of his own family. The guest's father had had my friend's uncle killed, and then been killed himself by the father of the bedouin who now sat beside him in the firelight.

These expeditions return the modern Arab, even if briefly and artificially, to the world of his forefathers, a world from which the responsibilities and preoccupations of riches and statehood have now made him an exile. The elaborate conventions of the palace, the headaches of defence and trade, are replaced by the more direct and familiar issues of the tribe and quarry. Perhaps the almost military resources of the larger hawking parties seem to us incongruous with the spirit of a bedouin holiday. But the changes of the last generations have been so rapid that the Arabs have not yet become self-conscious or precious about their origins. To ride when there is a jeep, or to change the cool boxes for goat skins would strike them as a breach of common sense. Thank God for electricity and clean water, for metalled roads and medicine, must be the honest and practical attitude of people who are no strangers to real hardship.

Though that may be so, returning to the desert with the accessories and armament of the factory, the human predator irrevocably damages the balance of chance and blunts the edge of sport. It must be admitted that some of the larger expeditions seem to catch the spirit of warfare more than that of the hunt. It is a natural if sad development. The predator is only too aware of the chances of an off day and so has an instinctive inclination to greed. Changes in weather can suddenly affect the availability of prey and predator's ability to catch it, so he feeds up when the going is good, as an insurance. A fox in a chicken run does not set the artificiality of the situation against this impulse. Nor are we so different. The thrill of hitting a high bird is the climax of the shooting man's skill and ambition, but we see nothing strange in the excitement of stories of days when 'the barrels were too hot to hold'. Wilfred Thesiger recalls that in the late forties, when he was hawking on the edge of the Empty Quarter, a daily bag of six or eight houbara was considered well done. By the sixties, this number had fallen off a little, but Roger Upton was still able to find regular flights when he was taken over the same ground by bedouin falconers. In the seventies, I found these places empty save for a few scattered remnants of the old stock. The sheikhs have therefore sought out new hunting grounds and the satisfaction of a few well caught is now giving place to a quest for higher and higher scores. Seven hundred houbara killed in ten days was the hair-raising report which came back from one party in northern India in 1976.

My own happiest memories go back to small forays with one or two friends when we could be sure that those which got away, did so to safety and not into the sights of another column of a large expedition. Some pieces from a notebook may give something of the haphazard adventure of these trips.

(After long promises that I would be allowed to join two friends on a planned hawking party, I was bid report to Buraimi Oasis at a day's notice.)

Friday: My optimism in expecting that I would get here in the midst of preparations for our departure is defeated. Mubarak was not even here when I arrived yesterday midday. He is apparently in Dubai, collecting some more hawks from the Al Maktum. His brother, Saif, is in charge of the house in his absence, and this morning made me a splendid breakfast of *khabisah*, a thick paste of flour and clarified butter, slightly heated to make the fat rise to the surface. It is eaten with strips of bread torn from a roll of layers, much the size and consistency of a folded airmail newspaper. A visit to the zoo in the morning; how young bedouin have great dignity when dealing with strangers. Two of Mubarak's sons came with me, aged eight and ten, and sat very solemnly with their feet dangling from their chairs in the director's office. They stirred their tea with studied indifference to all the exciting things like skulls and eggs which littered his room. The tension broke after about half an hour and they slipped out to charge all over the zoo, throwing stones at the wolves and tigers, and pressing their noses to the glass panes in the Snake House.

Afterwards, I said I wanted to buy some more pigeons as hawk food for the trip, so we toured the *suq*, or market, but without success, and then went to Sheikh Khalifah's house to try to sponge some from him. We wound up empty handed in an area where the Baluchis live. Here we managed to buy a squab for five dirhams, only to find on our return home that some child had let two of the three which I had brought from Abu Dhabi out of their box.

Just before sunset, I flew 'Bitch', my new haggard peregrine, loose to the lure for the first time. Great amusement and excitement from the children and others when the lure line slipped from my hands as she hit the lure and carried if off to a nearby sand dune.
Saturday: Mubarak has at last turned up. But, with the indifference to time so trying to foreigners who live by programmes, suggests that we leave tomorrow morning. I must see if I cannot get them round to setting off after lunch. If it will take eight hours to get to the Jiddat al Harasis, then we may as well make a start today. Vain hope. Off shopping for supplies: a huge sack of flour, onions and, very tactfully, some Enterovioform pills. First of all to the bank where we were received like sheikhs by the Pakistani manager, and took out £400. Shopping was therefore a rather carefree business and I was constantly nervous that he was on the brink of committing some horrifying extravagance.
Early evening: After lunch, to sleep curled up in my *farwah*, a fleece-lined cloak used by northern bedouin, in a corner of the *majlis*. I was woken by 'Bitch' whom I had brought inside because of the wind. She had got her hood off and was bating. I sat up to find myself being stared at by eight strange Arabs, most of them apparently well on in their hundreds, seated round the wall of the *majlis*. I stared back in confusion, hooded the hawk and scrambled out of the cloak. Then I recognized one of them as coming from al-Wagan, an encampment ninety kilometres south in the sands. They were a party of Duru', a tribe which lives on the eastern edge of the sands between 'Ibri, Umm al-Zamul and Nazwa.

Mubarak was dragged from his bed by the Baluchi servant – then great coffee drinking and date eating. They seem to be staying and staying (I later learned that the delicate matter of their visit was to seek support for an application for UAE (United Arab Emirates) passports) and so I have retreated against the wall to write. The South Arabian habit of standing up whenever anyone joins the circle is hard on the knees. In the north, only a few get up and only for their seniors. Slightly absurdly, one of Saif's sons who is six came in and at once we all had to stand to greet him. A little later his brother, aged one and a half, also came and then the same performance, except that someone carried him round the *majlis* for the ritual greeting of pressing noses. Some of the Duru' have fantastic Old Testament beards, but there is no trace of a smile as they push their chins out to touch noses. They hold their fingers out to the chest of the welcomed as if ready to resist a sudden access of emotion.
7 pm: Apparently Mubarak will have to kill for them which seems something of a burden, with goats and sheep going for six or seven hundred dirhams (£75–£90). A complete carcase from a butcher costs less than a third of this,

being deep frozen and imported, but such an offering would be an appalling insult. This was all explained to me while we had new sand tyres fitted to the Toyota pickup.

Monday lunch time: Dar al-Wahibah. It seems to me that unless we kill, there is no time for writing. Yesterday we found nothing but today Mubarak's hawk killed a houbara, so lunch will be a protracted break.

We left al'Ain at eight in the morning and after innumerable stops for, amongst other things, a hundred and twenty-three gallons of petrol costing 227 dirhams, (approximately £30), and some drinking water from an irrigation canal, we got to Qabil and the Omani border. Mubarak's rifle and shotgun were impounded by the customs officer against collection on our return. Thence to 'Ibri where we bought in that noisome *suq* a huge bag of dates in a woven basket of palm leaves. This was a triumph of Palm Sunday origami: it appeared not to have any seams, openings or closings and weighed ten kilograms (over twenty pounds).

From 'Ibri we drove south for three hours, stopping on the way for lunch of a handful of dates and a little coffee. Later, we caught up with a family of Duru', with whom we had more dates and coffee. They told us that there had been no rain on the Jiddat al-Harasis, and therefore there were 'no Arabs nor living thing on it'. After that news, we bore east to come here to the Dar al-Wahibah, a vast rolling gravelly plain cut by winding dry watercourses with thorns and shrubs. *Dar* really means house but can also mean realm; hence *Dar es-Salaam* (of peace) and *Dar al-Harb* (the *dar* of war) by which Muslims mean any and everywhere outside the lands of Islam. Here, the bedouin use *dar* to mean their stamping grounds. The Wahibah tribe lives in a band across this plain which stretches eastwards, below Nazwa, to the Wahibah Sands, southwest of Muscat. Scattered among them are the Harasis who gave their name to the Jiddah, or highway, which we have forsaken.

In the evening, I saw a hare in the distance, but it got into the bushes in a wadi and we never found it, nor anything else. There was not a breath of wind and a clear sunset. We made a stew from some onions and rice and I slept in my *farwah* under a wide sweep of limpid stars.

This morning's flight started with some tracks in the soft earth on the edge of a shallow wadi. Mubarak spotted them from the car, and we followed them on foot into the watercourse. As we were working along the bushes, a bustard broke cover some thirty yards ahead of us and after a few quick steps took flight. Mubarak slipped his saker and she tore after it as it made towards the top of a rise ahead. With the hawk well up to it, the houbara put in and at once fanned out its tail. With an agitated quick step it kept its stern towards the hawk and rattled its wings. The saker landed and with a few gangling steps hopped and jumped at the houbara, but came down against the shimmering tail. A second leap caught the houbara while turning, and she had it by the head in a flash as we ran up the slope. The bedouin love this sort of thing. They admire the hawk's courage for attacking the houbara on the ground, certainly a David and Goliath affair.

Tuesday: Under a *sidr* tree in a tributary of Wadi Sidiriatain: a long way east of yesterday. After lunch, we drove east through well-watered wadis. There must have been some tremendous rains, some watermarks were six feet up the banks. Mubarak said that the pools left over were the remains of two months' standing water. In going eastwards we were looking for a friend of Mubarak's, Salim bin Saif, who would be able to tell us definite news about game. He has been wandering the plains with his camels for a couple of months since Mubarak last saw him. We are virtually at the end of the season and fear we shall have difficulty finding flights.

While looking for Salim, we asked everyone we saw for news of him – a man riding alone on his camel, a man and a boy riding and leading two more. They were heavily armed and had very cavalier streamers falling from their saddles. We came across an old man with a pair of field glasses who simply appeared to be at a loose end; and, of course, a fair number of women. With their men off earning money in one of the armies, they are left to herd the goats and camels. The women are superb. None of them much over five feet, they are swathed in black gauze with enormous black masks. Tiny dark green sleeves and leggings stick out of this bundle while the only visible flesh – thin, bird-like hands framed in heavy silver jewellery – gesticulates violently as though conducting their shrieky voices. One also notices their miniature black feet which they slip out of their sandals as a token of respect when talking to a man.

We had a couple of flights at hares, but they were long slips and both of them found refuge in foxes' holes well ahead of the hawks.

Thursday am: Wadi al-'Ain, on a parallel with 'Ibri. On Tuesday night, we camped with a family of Duru' whom we had seen on our way south. They agreed that the summer was not far away. The south is already warming up and the houbara will be moving on fast, if they have not already done so. These Duru' had an attractive pair of salukis which were very en famille. I made friends with one of these and when we had eaten, which as guests we did first, I suggested that the salukis have their turn. This turned out to be an incomprehensible joke which did not go down very well. There are innumerable little footfalls in manners and politeness which in all honesty one could

OPPOSITE PAGE *Imping*

not possibly foresee, but which, with a gruesome logic, are explicable once one has recognized them. There are also commonsense points which one only learns by experience. Do not be the first to leave the meal which is eaten from a common plate. I, at any rate, find it very hard to distinguish between the two bowls of water which lie to one side, and often find myself being asked not to drink from the washing bowl or vice versa. Let them sort it out and follow suit. When being told to come and eat, do not go up-wind of the food: one is bound to kick sand into it. The sand which God and the wind have already put there is one thing, sand off a foot is another.

Yesterday there was a bad sand storm which we suffered the more for being surrounded by dunes, having come into the west and off the gravel plains. Nonetheless, we found a large bush and camped down behind it. Mubarak said no, I could not skin the hare which we were going to have for lunch; the given reason being that he was going to say his prayers, but the anterior one, I suspect, was that he did not believe I could do it without messing up the lunch. A little irritated with this assumption, of my own as well as Mubarak's, I let them get started with prayers and then just did the job which caused some amusement. al-Ra'aish was particularly amused.

al-Ra'aish is the third of our party. He claims to be related to Mubarak in some way and indeed may be, as he too is a Mansuri. But he is certainly not a brother from the way Mubarak treats him. He rides, with his rather sleepy saker, on the back of the pickup along with all the luggage and the kid which we bought from a Wahibah girl. While we were getting out the cash she stood eyeing us from behind her mask and impressed us by the way she held her hands on her narrow hips. She was delightfully oblivious of the kid's struggles as she held it down by a hind-leg gripped between her toes.

al-Ra'aish has beautiful manners and an equally beautiful, if unexpected, blue pinstripe jacket which fits superlatively. Covered in mutes and sand, at forty-five or more, he cuts an improbably distinguished figure. In some ways he is a sort of bedouin gentleman's gentleman and has all the appropriate disdain for those whom he regards as being of lesser blood than himself. After all, he comes from a deep sand tribe, the Manasir (singular Mansuri) which is based on Liwa and dealt in the past mainly with camels, and that gives him great snob appeal. He has also been hawking in Pakistan, by aeroplane, a fact which deftly recurs in his conversation. He adores talk and when Mubarak is not about, regales whatever company there is with long and complicated stories about camels and hawks. He has a strong sense of humour and keeps himself amused on the back of the Toyota by repeating to himself cherished lines and aphorisms.

al-Ra'aish is an expert at *dhub* catching. A *dhub* is a lizard about a foot long with a heavy, wide tail and a broad malignant toad face. The bedouin say they are good eating and whenever we see one, we leap out of the car to catch it – even if it means ten minutes' digging down a hole. I refuse to touch them but beat them out of bushes for al-Ra'aish. Their skin is like fine rubber and squeaks as it is pulled off. There is a surprising lot of blood in a *dhub* and the hawks do very well on them.

After our lunch in the dunes, with ears and eyes full of sand, we drove to a friend's camp and holed up there for the rest of the day. We hoped that the storm would blow itself out during the night. They had an old white canvas army tent which was sandproof in some corners. Fortunately they were a talkative and amusing lot and the hours till sunset passed quite quickly.

To be told to sit tight from two o'clock one afternoon until the middle of the next morning would be a tall order at home, but the impossibility of doing anything else makes the sitting bearable in the desert. For supper we had some of that dried fish which smells so strongly when on sale in the *suq*. When boiled up with rice into a sort of kedgeree it is not too bad.

This morning one of the bedouin who had crept into the tent yesterday and said that he had seen a houbara on his way over, led us out onto the plains to show us the place. After a couple of hours spent combing the wadi beds, we found some fresh tracks and followed them until we found the brush marks made by the bird's wings as it takes off. So we drove to a rise in the land and held out Mubarak's hawk, unhooded, to see if she could spot anything. After gazing about in an imperious way and a couple of rouses, her eye caught on something and she started to bob her head – an infallible sign. Mubarak let her go. With a couple of slips she was high up in the strong wind and a second later, some quarter of a mile away, two houbara jumped up out of the gravel. They went down the wind, wings beating in slow, deceptively slow, unison, the hawk sculling along for all she was worth to make up with them. We watched them out of sight with field glasses, then, when the last of us had admitted to losing them, we leapt into and onto the Toyota and bounced after them.

Coming to a rise, we saw below the instant tell tale of a kill – a train of white feathers blown along by the wind, sticking shivering on the stones, and at the end of this – the hawk on the dead bustard. Great relief and lots of cigarettes.

At the end of the season, with the departure of her quarry for the north, the hawk's work is done and the falconer releases her to the open sky whence he took her. The hawks spread their wings and depart, hopefully none the worse for their brief liaison with the Arabs.

The western mind often starts at this apparent prodigality. Hundreds of thousands of dollars are spent on these hawks and then they are simply turned out; our materialist standards of thrift are outraged. Furthermore, the hawk's release passes without any ceremony. At the end of the day, their furniture is removed and they are slipped down the wind and left to their own resources. There are none of the fanfares and deliberations which would perhaps attend such an occasion in Europe. Such circumstances are foreign to the pragmatic desert mind and the emotionalism latent in these ceremonies is absent from the Arab falconer's relationship with his hawk. In our own world, we can hear a sure but distant echo of this attitude in a terse exchange in Aelfric's Colloquy,* a schoolroom reader dating from the tenth century. The conversation, drawn, presumably from familiar subjects, is set out in Latin with an Anglo-Saxon crib to help the pupils. This passage exactly sums up the bedouin's attitude. The Master speaks to the Falconer:

'How do you feed your hawks?'

Falconer: 'They feed themselves and me too in the winter. In the spring I release them to a wood and, come the autumn, catch up the young and train those.'

Master: 'And why do you let them go when they are trained?'

Falconer: 'Because I do not want to feed them in the summer – they eat far too much.'

Master: 'But many people keep their trained hawks through the summer and then have them ready again.'

Falconer: 'So they may do, but I do not want to have so much work with them (my hawks) as I know how to catch others – and not just one, but lots.'

*Aelfric (A.D. 955–1010) English abbot and author, a great prose writer of Anglo-Saxon times.

OPPOSITE PAGE *A falconer mounted on a fine thoroughbred riding camel carries a haggard peregrine.*

PAGE 102, ABOVE *Passage peregrine with a desert hare she has just taken. Many peregrines are reluctant to fly hares as they are nervous of a fight on the ground–they much prefer to attack winged quarry in the air.*
BELOW *A young peregrine spreading her wings to steady herself on a houbara kill, taking hold of the head is the only way to deal with such large quarry.*

PAGE 103 *A passage peregrine beginning to pluck her kill, a stone curlew.*

PAGE 104, ABOVE *A falconer helps a passage peregrine feed from the brains of her kill.*
BELOW *Salukis and a team of trained hawks. Note how the saker's expression and fluffed breast feathers show her relaxed tameness. Only two of the four peregrines behind her show the same restful state.*

Home Life

'He found him in a desert land . . . and kept him as the apple of his eye.'
Deuteronomy.

EARLY BOOKS ON HAWKING NEARLY ALL HAVE LENGTHY SECTIONS devoted to illnesses and mishaps, and offer the reader great lists of potions and powders which will hopefully remedy them. For a falconer in retrospect this is scarcely surprising as the problems of disease are daunting. Indeed, to train a hawk and fly her in the Arab fashion is no great difficulty; there are few intricacies of refined technique. All that is required is a fit and sharp-set hawk and this, for brevity and clarity, is what all chapters on training and flying assume as read.

I have often thought that by comparison with the creatures of other sports, hawks are in some respects straightforward to keep. They do not, like some horses, waste hours of their master's time refusing to be caught before the game has even begun, nor, like dogs, do they cringe and whine for daily and strenuous exercise. They make few or no emotional demands and are content to be left in peace for much of the day. They are kept tied up (so the falconer knows where they are, or assumes he does), and deriving their requirement of liquid from their food, they do not need constant access to bowls or buckets of water for him to fill up and knock over. They are, however, their falconer's hostages and depend entirely on his care and attention. High scores reflect, as much as anything else, the determination and energy of the falconer in finding slips for his hawk. A surer guide to his ability as a falconer is the condition in which the hawk achieves her success.

For the Arab the problems of management are limited by the fact that most of his hawks will only be in captivity through the winter, and so the proportion of days at home to days in the field is much smaller than in Europe where they are kept from year to year and must sit out the close season at home. Nonetheless, there are health problems.

Being a bird of prey, the hawk, of course, exists on fresh meat and nothing can substitute for this staple. But in eating her quarry she also takes in a variety of essential vitamins and minerals from the innards and bones of the bird or animal she has killed. Bones and skin, whether feathered or furred, give her the roughage which she needs to clean her stomach and crop, and the passing up of a hard pellet, or casting (*rubajah*), composed of the indigestible remains of this roughage, is a bodily function vital to her health. A cut of butcher's meat falls far short of the nutritional value of the all-in food parcel which she is accustomed to catch for herself in the wild. But her natural food is not always available to the falconer and then she must make do with mutton or goat. This is no bad thing during her training: the thinness of the diet will break down some of the wild bird's reserves of fat and this must be done to make her amenable and obedient. By the time the hawk is honed into flying shape, it will not be long before she is getting a share of kills in the field. This is particularly important for the peregrine which needs a higher quality diet than the saker. Sakers and lanners seem to do well enough on coarse meat for long periods, in the same way as goshawks do outside Arabia.

The flesh of different animals and birds varies considerably in food value and, as a hungry hawk makes a good gauge of this, the bedouin have a sound empirical knowledge of the differences. A sharp-set hawk in flying condition should be fed the amount of food which will bring her to exactly the same state twenty four hours later. So enough mutton to give a saker a crop about the size of a large egg, would make a reasonable daily allowance. The same amount of pigeon would probably raise her condition, likewise hare. But that quantity of chicken would be insufficient, and she would be very hungry well ahead of her next feeding time. Camel's meat is considered very heavy for hawks and they take a long time to 'put over' their meal from crop to stomach when they are fed this. In practice, the basis of the hawk's diet is inevitably sheep or goat, but as the bedouin reckon that on no account should a hawk go more than a week without casting, this is punctuated by regular pigeons. Children may snare small birds or kill a few with their slings, but these are only occasional offerings and cannot be relied on. In a case of real need, a short ride into the desert night with a car-load of children will usually bag a fair number of jerboas which the children course in the headlights with much shrieking and colliding. Though this is something of a last resort, sakers love this food and quickly put on weight with it.

Finding enough of the right food is always a headache and certainly an expense. It is a bedouin custom that if a household is killing a sheep or a goat, and they know that a neighbour or somebody in the encampment has a hawk, they should send a cropful of the best meat to the falconer to help him out. Further, a guest for whom something will usually be killed out of hospitality, may ask for meat for his hawk without any embarrassment.

Within that broad outline of diet, there are other more particular concerns. Hawks, like ourselves, have individual appetites and metabolic rates. Roughly speaking, small hawks eat a good deal more in relation to their size than do large ones. Peregrines tend to eat more than sakers. So a peregrine tiercel will need more careful dieting than a large *hurr*, or female saker. However, the falconer must discover by trial, but, especially in the case of small hawks, not too much error, what is each bird's own requirement.

This in turn is affected by the weather. In Europe, small garden birds, we are assured by the makers of food sold for them, can lose up to a third of their body weight keeping themselves warm during a winter night. That extreme does not apply to hawks in the desert, but it illustrates the nature of the problem. An early frost – and after Christmas the desert can be extremely cold – may catch the falconer out, and he finds the hawk which he was keeping sharp and ready to be flown is more in need of immediate food than exercise. A hawk which is below par in this way very rarely gives a good account of herself in the field. The ideal is to find that level of feeding which keeps the hawk in a buoyant state of aggression but not so high that her obedience is put at risk. Arab falconers assess their hawks' condition by feeling the flesh on either side of the top of the breastbone and at the bottom of the chest by the legs.

If the feeding problem can be mastered, the falconer has gone much of the way to prevent any illness or disease breaking out, as it surely will if the bird's defences are lowered by a reduction in condition. Externally, the hawk always runs the risk of injury, principally to her feathers, if her block is not properly sited.

A broken wing-feather, apart from being unsightly, seriously reduces her speed – as anybody who has tried to swim with his fists clenched will know. Similarly, a break in her tail affects her manoeuvrability. Another anxiety is that one break very often leads to more: a bird's wing or tail feathers overlap each other and in doing so give each other support. A broken feather leaves a gap and the two on either side of this become more accident prone. With alarming speed the whole is ruined. Fractures and breaks in feathers must therefore be repaired with all haste. A badly bent feather can be repaired and straightened with very hot water. The hawk is hooded or held by a friend and the water poured slowly down the damaged quill. The falconer rolls the feather very gently between finger and thumb and moulds it to its proper shape. This treatment, however, tends to make the quill brittle and, if the fracture is serious, it may well be safer to cut it clean and proceed as though it had broken completely.

This is a painless operation as the feathers have no nerves and is called 'imping' (*tawsir*) and consists simply of rejoining the broken piece by means of a pin inserted into the quill. A spine of metal, bone or horn, is pushed into the detached quill and the glue on it allowed to set dry. The sharp point protruding is then put into the stump and pushed home: the idea being the same as the surgeon's intra-medullary pin. To secure the joint further, some falconers splice on a piece of quill taken from the feather of a chicken or pigeon.

More often than not broken feathers are the result of repeated bating from the block. The wings are scraped against the ground and the leash, running from the hawk's legs back to the block, passes through and abrades the tail feathers. Manning and careful use of the hood are the only prevention against this. Some European falconers prefer to keep their hawks in a state of near wildness so that they can only be approached or flown by their master and therefore stand less chance of falling into the wrong hands or being shot if they are lost. However, this option is not open to the bedouin as explained in the chapter on training. Privacy is unknown in a *majlis* and the hawk is unlikely to be kept among the women where it may be molested by children. So the hawk must quickly reach a high standard of tameness and be content to sit quietly amongst the men. There is some advantage here in that the hawk is always under observation and any mishap, such as a tangled leash, can be quickly put right.

A final aid to good health and feathers is to allow a hawk to bathe from time to time. Although hawks do not drink and have no imperative need of water, they do, with the exception of most sakers, enjoy bathing. Getting thoroughly wet encourages them to make a proper job of cleaning, oiling and ordering their feathers. In very hot weather they may take a few sips to drink, especially if they have been fed on butcher's meat which tends to be dry. The bedouin also have a ruse to settle down an agitated and nervous

OPPOSITE PAGE, ABOVE *Houbara caught alive at the beginning of the season are often used to train young hawks. This presentation of prey to hawk for the first time is known as 'entering'.*

BELOW *Three sakers sit on a cadge in the early morning sunlight waiting for the day's hunting. They are hooded but remain untied, and will rest undisturbed until the hunting begins.*

hawk which is to take a mouthful of water and spray it through pursed lips in the hawk's face. The hawk usually opens and shuts its beak to take in some of the water and finds relief in this. She will also, once wet, be less inclined to bate – a fact noticed by old falconers in the West who used to drench a hawk which was hood shy and being obstinate.

If these circumstances and conditions are attended to, there is a good chance that the hawk will remain healthy and in good fettle. She stands on her block with feathers half fluffed out and one foot drawn up in relaxation. Yet she is alert and has a bright eye which takes notice of everything. From time to time she lowers the drawn up foot and rouses – lifts her feathers till she looks more like a feather duster than a bird of prey, then leaning slightly forward, shakes herself vigorously to bring every feather down in its place. This may be followed by a 'mantle', stretching the left or right leg out at an angle under her tail, and drawing the same wing over the top of her fanned out tail to comb the web of the primaries. After this, she 'warbles' by lifting her shoulders up and extending her wings to stretch, spreading out her tail feathers at the same time.

These perfunctory exercises may tempt her to set about a thorough servicing of her feathers, preening and digging, combing and pulling, returning frequently to the gland at the base of the tail to pick up oil on her beak and renew the bloom on her feathers. Another rouse – this is a good sign of a sound hawk. But the sight of her falconer approaching with food suddenly transforms this domestic figure into a hard-eyed hunter, sickle wings and grappling iron feet. She takes to her food with energy and no hesitation. When finished, she runs her beak round the edge of her block to clean it and puts over her crop, converting it into clear black and white mutes, or droppings, in about six hours.

Any departure or omission from these habits should start the falconer wondering what may be amiss. A glazed eye, feathers erect but prickly and immobile, pasty or congested mutes, flicking away her meat as she pulls at it: such signs spell trouble. But the bedouin, himself a survivor and not an innovator, is ill-equipped to cope with serious disease. He compensates for this in part by his keen powers of observation and very highly developed intuition in matters concerning animals*, and if he reacts quickly, he may

*When I was living in Jordan, my riding camel developed swollen ankles and so I pestered a vet from the Ministry of Agriculture to pay her a visit. He kindly did this but had to admit that my best course was to seek advice from the bedouin. They, he said, knew more about camels than all the vets in the kingdom. Disappointed of a scientific treatment, I did as he suggested and the camel soon recovered. The bedouin's immediate (and I have to admit, correct) diagnosis was that I had been riding her too hard. A rest was all she needed.

ward off trouble by feeding up his hawk and giving her a rest. He is helped by the fact that the desert climate and environment are essentially healthy. There is little water to help germs or bacteria grow and spread; the strong sunlight has a sterilizing effect and nomadism in fact results in highter standards of hygiene than, regrettably, one might guess from the state of modern settlements. The sentimentalist would at least have to admit that man and hawk are in the same boat – how many of us would care to live in our European mews and hawk houses?

But when faced with a serious disorder, the bedouin must turn to rudimentary palliatives and remedies of the kind he would just as likely apply to his own family. It would take us far from the scope of this book scientifically to describe the many diseases to which hawks are prone, as only a few of them are recognized in Arabia. Those interested in hawk medicine in general can do no better than John Cooper's *Veterinary Aspects of Captive Birds of Prey* (Standfast Sporting Library, 1978) which draws together and explains the present state of knowledge in this field. Anyone hoping to find in these pages curious and recherché remedies will, I am afraid, be disappointed – as I have been when, separated from my books, I have sought help from tribesmen about various ailments. Where their advice has been possible (I have always shied away from the resort of branding in which they have so much faith), I have found it simple and sensible.

Lesions in the mouth (frounce) and on the skin (pox) are two distinct disorders but both known by the same name in Arabic, *jidri*, though the first condition is often fatal and the second is unlikely to be. The sores are painted with gentian, or, externally, branded. Bumble foot and corns (*sidd*) are similarly treated though optimistically, as the Arabs profess to know of no cure for these. The hawk does not die, except in extreme cases, but loses the use of her claws, or even the whole foot, and so becomes unserviceable to the falconer. Blains and blisters are lanced and washed. Internal disorders (constipation, crop inflammations, loss of appetite, badly discoloured mutes and, when apparent, worms) are all treated with an immediate change to a lighter diet, preferably of lamb or mutton if available. This may be soaked or dipped in milk or egg yolk and the meals kept small and frequent. Sugar candy (*sukkar al nabat*) is sometimes used for constipation. All of this is sound and plain advice but skims quickly across the surface of the deep pond of illnesses and diseases into which captive hawks may sink.

The difficulty in making hawks take any liquids and their quick aversion to any strange taste, effectively rules out the products of the bedouin amateur apothecary. Though they do make herbal teas for themselves, I have not come across any such prescriptions for hawks.

Although its use is not indicated by a disease, mention can be made here of dosing hawks with small pills of sal ammoniac (*nashadir*) to bring them quickly into flying condition. In the West, falconers accelerate the reduction of a hawk's high condition by feeding her beef which has been soaked in water to extract its goodness. This practice, however, is not used or known among the Arabs. In warm weather, when a hawk's food requirement is very little, they face a problem in how to get a hawk sharp without a period of days on very short commons which will also weaken her. Their answer is to force feed her a small dose of sal ammoniac. The hawk very shortly casts this up and with it a fair amount of internal grease, leaving her extremely hungry but not debilitated by fasting. Needless to say, this drastic measure can only be used with safety in reliable weather and on a hawk which is absolutely healthy.

Only too often, a sick hawk is in no time a dead one. A hawk has a great capacity for carrying a disease without showing its symptoms, and, by the time these appear, the illness may be far advanced and beyond any treatment but by the most skilled use of modern drugs, unavailable in any case to the bedouin. They take a detached view of death which can seem hard to the foreigner: 'God's we are and to God we return' and 'No man knows in what land death will find him' are sayings which reflect their attitude and are constantly on their lips. An extension of this resignation is that they will not put down evidently sick and helpless animals. Their argument is that away from the hunting field it is for God to take the life which he has given. So a camel broken down with exhaustion on a long journey is turned loose to fend for itself, even in an area where all know there is no feed or water, no matter how much the meat would be welcome to the riders if the animal were slaughtered. Healthy animals who come to grief slipping on wet mudflats (camels can split themselves in this way) are left for hyenas and others, rather than killed and the meat saved. So with their sick hawks, there is an element of fatalism in the falconer's equanimity as the bird's condition deteriorates. The best of them will release the patient to try her luck on her own.

From such gloomy prospects for the hawks which fall below the line, let us return to the end of the season in March when the falconer releases his team 'down the wind, to prey at fortune' and breed a new generation of passagers for the next year. If he has the money and facilities, he may well decide to keep on the star performers and put them by for the summer. During the hot weather they will moult their feathers and the tax on their constitutions of producing new ones is considerable. They must be kept in a very high condition to enable them to moult properly. This in turn means that they become unmanageable and must be left in peace. Their owners must have a room or loose box to house them and somebody available to see to their daily feeding. Hawks can be moulted satisfactorily on their blocks, provided that they are undisturbed and, if they bate, they can land on a soft surface such as sand. Otherwise, they can be released in a loose box. I know Arabs who have tried both methods and they find little to choose between them. An interesting point is that while they appreciate that moulting hawks must have somewhere undisturbed, they do not extend this to those in training who must live with the men and keep their hours.

In 1975 Sheikh Zaid, the ruler of Abu Dhabi, very kindly invited me to moult a peregrine I had at that time with his own hawks. He told me that the summer humidity on the coast was very hard on hawks and that therefore he had set up a mews about fifty miles inland where it was much drier. There he had had two large sheds built from woven palm leaves and employed a group of Pathans to guard the twenty-one hawks. They were visited daily by a tribesman who supervised the killing and dividing of a goat, all of which was devoured by the birds. Twice a week, pigeons were taken from the dovecote and each hawk was given half one of these to allow her some casting and variety. Evidently this was a very expensive operation of a kind which would have been very hard to manage thirty years before.

As did a contributor to *The Boke of St Albans*, the first printed text book in English on falconry dated 1486, the Arabs advise a dramatic change of diet if a hawk is slow to shed her feathers. A newborn kitten was suggested, but mercifully not obtained, for my hawk staying with Sheikh Zaid's. Sakers should be through their moult by October which is about as early as the falconer can have any use for them. Peregrines are usually slower to begin and thus end later. When they are taken up, newly moulted hawks are unruly and fat. They must be reduced and put through a course of retraining. During the first few days they are still very soft and unfit. In this fragile state it is easy to ruin them, perhaps even kill them by bringing on an epileptiform fit or an attack of hyperglycaemia as a result of clumsy handling. But happily they remember their training and are soon returned to their former selves, if this initial hurdle is carefully negotiated.

A frivolous footnote here is the fact that moulting hawks sometimes spontaneously lay eggs, evidently sterile in the absence of a tiercel. Telling the sex of hawks was an interminable source of conversation with some very old friends in North Arabia. In an intensely masculine society, the supposition that the biggest and strongest must be male was not only attractive but obligatory. When I adduced the argument of the eggs with one falconer whose hawk had laid during the moult, he considered my remark most tasteless and out of order.

End Piece

' "Once a falconer always a falconer" is a maxim of universal truth.'
E. B. Michel.

Thus have they flown and thus do the Arabs still fly their hawks. Despite the great changes which have run through Arabia, the style and techniques of the Arabs' hawking have largely remained the same. It is the scope and effect of their hunting which are new.

It is easy to fall into the temptation of believing that our own episode in history is particularly significant; but in Arabia this thought is perhaps justified. The assault of modernity on isolated traditionalism has now made ground after some early skirmishes. Certainly those generations living into and through the last quarter of this century will see changes which will make them feel as remote from their grandfathers as we do from our own ancestors who lived before the industrial revolution. I do not mean that hawking will disappear or that camel herding will vanish as a way of life, but these things express man's relationship with his world and it is that relationship which is changing irrevocably. The Arab of today lives in contrast with an environment where others within memory used to survive in harmony. They brought little new to it and left little trace behind.

In the past when men left the desert and the nomadic life, they joined a settled society which viewed the bedouin with hostility. The open 'whiteness', the bedouin's home, was a separate world which went its own ungoverned way protected by the immunity of its remoteness. But oil in the ground, in the sea, in pipelines and motor cars has ended that. The bedouin have not ceased to be bedouin because they have opted out of the desert, or because they have been expelled by stronger tribes from within and then moved across the cultural barrier which divides the desert from the sown. They are as actors in a film which suddenly stops on a particular frame. For want of the possibility of continuing, they have to descend from the screen and join the world of the audience. They are no longer exactly the figures in the film, as that setting has no motion or independent life, but then neither are they quite as the audience. They rehearse their screen roles as far as that is possible, but are inevitably affected by contact with the audience and become increasingly self-conscious. They are estranged from themselves but have no other ready part to play.

So the Arab will continue to fly hawks. But he will do so without the careless ferocity and zest of the past. Like us, creatures of our traditions, the Arab will in time go into the desert to see his hawk fly and not to quench a desire for meat. He may go to remind himself, or to prove to himself, that he is an Arab by seeking involvement in his own customs. But if he tells a foreigner to let his hawk be unhooded and 'watch the Arabs', he will mean something new.

This process of alienation is already under way. I remember well another birthday which I spent with the 'Awamir, a tribe of the Empty Quarter. They were holding a marriage feast. Around the fires lit in the sand dunes sat circles of white-clad tribesmen, many of them weekend visitors from the town where they had taken work. The young men danced in lines and chanted old songs full of the heavy assonance which they love. There were a few salukis and a great many camels, some fine ones brought in for the following morning's races. The camels moved with indifference among the parked jeeps, trucks, water tankers and the pickups full of oranges for the party, the good things of oil. Among the circles, talking and beaming into the firelight, were old men also dressed in white. On the remittances and subsidies from the capital they lived a happy retirement, the burden of nomadism lifted from their shoulders.

But there were others, camel men in worn robes of browns and russets without a stitch of white, who had spent their lives in the deep sands and would return there early the next morning, when the feasting was done. They were less at ease and for the most part silent. Some of these remembered Wilfred Thesiger's journeys quite clearly. Only seasons separated one unusual occurrence from another. They had never learned the new Arabic of radio broadcasts and political speeches and much had to be repeated or paraphrased to them before they understood. Their own language was a thick dialect full of descriptive words for different things which to us were indistinguishable. One or two of them, I remember, would not drink tea or coffee; they were content with water, and presumably saw no point in nursing new appetites. Perhaps that would have been a senseless breach of the self-denial which had lengthened their chances of survival.

There was talk by the fire of whether one of the sheikhs from Abu Dhabi would come by helicopter – a visit had been expected earlier in the day. Had he come, he would have found a familiar scene but one underpinned by artificiality and alien economics. The party, a tribe at celebration, had in a sense lost in years the cohesion of centuries.

In closer focus and with regard to desert hawking, these trends pose some critical questions for today. Firstly, will changes in attitude, though probably inevitable in the long term, be effected in time to prevent the destruction of the quarry species? In many places the three traditional quarry of the falconer now verge on the point of extinction. How can this be remedied? Some awareness of this problem has been shown by those in authority and in some countries, notably Saudi Arabia, stringent conservation laws have been brought in. Encouraging though this may be, the difficulty lies at the executive level and in the general population's readiness to acknowledge the point of the sanctions. The prohibition of shooting in Saudi Arabia was a bold step; but the news from hawking parties is bleak and suggests that some restriction on falconers' bags as well will be necessary. If the residual stocks are so low, what difference does it make if they are further depleted by the shotgun or trained hawk? But the imposition of bag limits demands the co-operation of hunters, if the measure is to be effective. All governments are reluctant to enact unenforceable legislation and in Arabia, where hawks now find astronomic prices, falconers tend to be the well-to-do and socially powerful; and this only strengthens that reluctance.

This dilemma is acute for some of the less rich countries whose territories encompass good hunting grounds. Those most attracted by the presence of quarry there are those whose financial aid is vital to the economy and stability of the state. Are some or all to be excluded and, if so, how? A Pakistani put it to me cynically by asking what price, expressed in barrels of oil, I would put on a dead bustard.

Partial success is possible. For some years, the hare has been protected in the United Arab Emirates and its population has made a significant recovery. But could the houbara then have its turn of immunity at the expense of the hare's return? In countries where it has not so far proved possible to make an accurate census of the human population, how could such decisions be made except on the basis of impressions, and how objective and disinterested would they be? The same problems attend any scheme to ban falconry for a period to allow quarry a respite and chance to restore its numbers.

If game cannot be satisfactorily protected or its hunting controlled, we should examine the possibility of artificially increasing the population. Here practical problems are compounded by those of science. Macqueen's bustard have never been bred in captivity. The great bustard has been bred, but not in significant numbers and the techniques used to achieve this success are not widely understood. It was an ambition of the zoo in al-'Ain, United Arab Emirates, to breed houbara, but so far they have failed to obtain a pair. The difficulties in getting a healthy and undamaged pair in Arabia may well be imagined. There are also psychological considerations to be explored. What effect will the trauma of capture and captivity have on the bird's reproductive disposition? These questions also apply to the stone curlew which shares the houbara's secretive and shy personality. If the riddle of captive breeding could be solved, those of release and returning the new stock to the wild state would remain.

The experience of those who have bred gazelle and oryx illustrates this point. After many false starts, captive breeding of these animals has become a fairly straightforward business of following certain management principles. Domestic herds have prospered and, in the case of the oryx, final extinction has been warded off. However, the demand for breeding stock placed a further tax on the wild population and the information of the bedouin of the Empty Quarter and western Oman, the oryx's last refuge, is that the creature is now gone. No proven solution has yet been established to the problem of returning captive bred animals to the wild in desert conditions though recently some Arabian oryx were sent from Phoenix Zoo, Arizona to form the nucleus of a semi-captive herd in Jordan, a country once included in the natural range of this elegant desert antelope. One problem has been that their techniques of survival have been passed from generation to generation, once this tradition is broken it is difficult to know the best means of educating the orphans.

Desert animals of necessity need enormous territories in which to find their living. In Qatar, the repeated outbreak of disease in a captive herd of oryx whenever its population reached a certain level, also indicated that they are mortally averse to crowding. In spite of the financial resources available, the fencing-in of huge tracts of desert to accommodate a few

OPPOSITE PAGE, ABOVE *Falconers 'making in' to a saker which has just killed a houbara. They must take care not to stand over the hawk and thus alarm her.*

BELOW *Falconers prepare the evening meal –a day's bag of houbara.*

animals in the hope that they would become independent would raise considerable, and probably paralysing, administrative problems. Experts and supervisors would have to be employed and looked after and, more importantly, given adequate authority. These conditions could only be realized by the sustained determination of those in power. No such scheme has yet become operational in Arabia. The Shaumari reserve, near Azraq in Jordan, may make a start and, hopefully, history.

But to return to the falconer's problems, the bustard, unlike the oryx, has wings and is in the habit of migrating; and there the analogy founders. The gloomy probability is that until the hawking parties which travel overseas for game find foreign stocks also running out, these delicate and expensive issues will not be grasped.

The conservation conscious reader outside the Middle East may well wonder why I have stressed the plight of the quarry, when everybody knows that most birds of prey, and all falconers' birds, are listed as endangered species. Does not the dwindling supply of hawks pose a more serious threat to hawking? The quick and logical riposte to that is that hawks are of no use to the serious falconer without quarry. It is true that hawks are now much more difficult to find than they were twenty years ago, and the stimulation which oil revenues have given to the market for them in the Middle East must have contributed to their decline in numbers. But in Arabia and Pakistan and Egypt, the latter two both new and now much favoured sources of hawks, the decline has not been as apparent as it has in other parts of the world. Arabia, straddling two major migration routes (see Trapping) is the staging-post for very dense aerial caravans of hawks from a broad summer scattering of populations coming from eastern Europe and western Asia. Only catastrophic reductions of the raptor populations of this vast area would be reflected in the narrow channels of Arabia through which they pass on migration.

Trappers, however, have noticed a steady fall in the supply of peregrines, but other falcons are still caught in fair numbers and obviously not all those seen are trapped. The flow of sakers, on the other hand, seems constant and if he has the money or opportunity to go trapping, the Arab falconer has no difficulty in getting himself a hawk. The vast majority of those taken are finally released or lost, so the long term strain of trapping on the migrant population is not as great as it would at first appear in the autumn.

Hawks, and especially the peregrine, have done well out of the development of conservation. Certainly the predator is especially vulnerable for being at the end of the food chain, but in more acid moments, I sometimes wonder if the hawk has not scored valuable points in the hearts of conservationists with her glamour, looks and élan. The more pedestrian houbara and its like are hard pressed to compete. There is an element of irony then in the fact that energetic efforts are now being made to interest the Arabs in setting up their own hawk breeding projects.

Hawks have been trained and flown for thousands of years, but only in the last fifteen or so has a breakthrough been made in Western experiments to breed them in captivity. The difficulties for the hawks are psychological. Advances only came when science identified the importance of the photoperiod in bringing birds into breeding condition and understood the need for a complete absence of disturbance. Once these requirements were codified, hawk breeding progressed rapidly and some breeders in the West have now achieved many successful seasons. Success, however, has mainly resulted from the pairing of birds taken as eyasses and evidently the products of their unions have been eyasses too. The experience of European falconers is that eyasses, if not brought to the field and quarry quickly, can be difficult to enter. There is some danger of the young hawk becoming dependent on the falconer for food.

The Arab falconer has virtually no experience of eyasses and he has not liked what he has seen. Falconers' eyasses outside Arabia are entered in August. By the beginning of the autumn, they are competent and successful hunters. In the Gulf or Saudi Arabia, these timings fall straight across the very worst part of the year when, with temperatures well over 38°C (100°F), and seething humidity near the coast, wild hawks are nowhere to be seen. The delay between fledging and the availability of quarry would result in the need for some very tricky entering. Many hawks might well fail to realise their natural potential and what would become of the tiercels bred in such projects? Very few Arab falconers would be prepared to take them on, as is indicated by their absence from the dealers' stocks. There would also be a scarcity of suitable quarry for the tiercels which refused hare and houbara. Furthermore, breeding schemes in America using stock taken from a migrating population have found that the young have a significantly reduced instinct to migrate. A great deal more research would need to be done on this behavioural aspect before eyasses could be released in Arabia.

These considerations, I think, should argue forcefully against the introduction of breeding projects to Arabia. They were driven home to me when I was present when a rich sheikh in the Gulf was given a captive bred peregrine tiercel by a visiting businessman. It was two years old and in perfect plumage, but it had only recently been properly trained and had never flown quarry. It was a screamer: that is it had transferred its dependence on its mother to its human owner and, suffering the immaturity which often mars unflown eyasses, still called to him for food. It was a frightful noise. The bedouin falconers and retainers of the sheikh were most amused by this toy-like creature, and fell to holding their noses to suppress their giggles when its handler demonstrated its dog-like obedience to the lure. Of one thing, however, they were absolutely sure: they had no possible use for it whatever. The present was refused, a pointed gesture in Arabia, and the ignorant and unembarrassed tiercel was carried away in disgrace.

Larger hawks, especially gyr falcons, who arrive in Arabia in similar circumstances are of course accepted. But those, unlike their donors, who stay behind to learn how they fare in the field soon hear that little or no good came of them.

The success of captive breeding has certainly brought great benefit to Western falconers. They can produce what eyasses they need, and, mindful of the ecological dangers of introducing alien species or races to new habitats, can hope to support the wild populations by releasing a few surplus birds. The case is different in the desert and I find it hard to believe that home-bred hawks will ever supplant passagers in the Arabs' estimation. That is not to say that they may not be glad to take on some eyasses from the West which have already been entered to quarry there in the autumn months before the desert season opens. But every falconer prefers to train his own hawk and this could prove a deterrent against such deals. Any responsible falconer sending a hawk to Arabia would wish to examine the question of the eyass's chances of migrating successfully. A concerted series of ringing experiments might help explore these areas of ignorance.

Looking ahead again to the future, another rewarding area of research would be to look at the resources of medieval Arabic works on hawking and hunting to establish a sounder system of nomenclature. The classical language has a wealth of names and terms for different birds, but today these are used haphazardly and without reference to the scientific advances made since they were coined. Time and the media may overcome regional and dialectical variations, but books and articles would benefit greatly from more standardization of usage. At the Abu Dhabi conference on falconry and conservation in 1976, there were recurrent difficulties of definition and translation. The French scholar François Viré has made great researches on the Arabic names of birds used in early works. Such research, allied to the practical understanding of today's Arabs, could pave the way to a more scientific basis on which to discuss conservation issues and plan field work. Such necessary projects would surely stand a better chance of success in Arabia if the Arabs saw that here was something which they could develop in their own language and idiom.

The problems of suggesting and presenting such schemes are great. It is so easy for us, concerned and anxious about the health of the environment, to appear patronizing and didactic in offering our experience to others. They naturally have different priorities and values. A pointed demonstration of this for me was the comment of a senior Arab who had been present at the conference in Abu Dhabi. 'Is it not shocking,' he said to me afterwards, 'that during this week of discussion, not one of the speakers has mentioned the many areas of human suffering? We have only been talking of birds and animals and the experts have been vying with each other for Arab money to finance their own private schemes and projects. It seems to me that conservation is just a commercial business like anything else.' The sense or justice of such a remark is not as important as the impression which spurred the Arab to voice it.

An indication of the scale of these cultural barriers is the Arabs' view of hawking in the West. I have already said how eccentric our training of small hawks, like the sparrowhawk and merlin, appears to the desert falconer. Unfortunately their impression of our game hawking is no less jaundiced. The waiting on flight of the falcon at grouse is regarded by many British falconers as the *ne plus ultra* of falconry. Dog, hawk and falconer must work together with great precision to overcome the unstable factors of weather and quarry. When successful, the peregrine's long stoop across the sky after a covey of grouse rocketing along the top of the heather is an unforgettable sight. So it seems to us. But Arabs who have been taken grouse hawking, have thought the whole performance facile and unfair. The hawk, they say, has too great an advantage in her towering above the moor; and the fact that very few of the grouse flushed are killed, suggests to them that the tactic is inefficient. What success there is, they tend to ascribe to the pointer and this they see as putting the hawk and falconer in a bad light. I need not rehearse the disagreeable thoughts which they entertain about any dog which is not *hurr* or a saluki.

Perhaps they found our delight in methodical techniques irksome and, with their black-and-white perception of values, found them silly for not being more effective. Their instinct for austerity, despite the shadow of rapaciousness which it throws, has no sympathy for our finely worked plumed hoods and tasselled gloves. Their own hawking furniture is ruthlessly functional and spare. One cannot but observe in Arabia that all the pretty hoods and gloves which are sent out as presents are never used. The richest oil sheikhs take out their hawks with the plainest possible equipment and would find any thought that they do otherwise faintly ludicrous.

Their hawking has the same spirit of pragmatism. In its worst manifestations, it has a blood-curdling carelessness for the damage it does to the delicate natural order. But we see it thus because we have had the advantage of gradual development, and it occurs because they have not. At its less extreme, hawking in the desert is the pastime of prince and tribesman without distinction. The rich sheikhs distribute the hawks they buy among their retainers and go with them into the desert and there regain the old fellowship of journeying and firelight which modern life is so quickly breaking down. If it disappears, they will not be able to repair or replace it with anything which they can call their own. The fortress door will have swung to behind them.

Appendix

FALCONRY IN THE MEDIEVAL ISLAMIC WORLD

This is not the place to describe in any detail the astonishingly rapid expansion of early Islam. It made of the revolutionary sect of its Arabian Prophet Mohammed a major civilization, reaching from the borders of China to the Atlantic and all in the space of the hundred and fifty years following the death of Mohammed in A.D. 632. The important point for us is that while the society which emerged had many features which were Arab, principally of course its religion and language, it was not Arabian. It was a brilliant, sophisticated amalgam of the Persian, Byzantine and other indigenous cultures which the Arab armies met in the course of their conquests.

Islamic civilization in its heyday was extraordinarily tolerant. It embraced and absorbed many new ideas and styles, changing them subtly and allowing itself to be changed by them in turn. The Muslim community was, in our own terms, cosmopolitan and international. To be a Muslim meant to be accepted throughout the territories of Islam, and more local differences of race or custom were of much less account set against this universal standard. The decline and fragmentation of the truly Islamic state followed its penetration by warlike and self-regarding outsiders from Central Asia who held their tribal identities above that of their adopted religion.

What we know of the history of hawking in those days follows this same pattern. Hawks were certainly being trained by the Arabs, probably in much the same way as they are today, before the time of Mohammed. There is a story in the *Hadith*, the collections of the Prophet's sayings, which records the conversion to Islam of Mohammed's uncle, Hamzah ibn 'Abd al-Mutalib, on his return from a hunting expedition, hawk on hand.

In their expansion abroad, the Arabs took their hawks with them. We know that Yazid ibn Muawiyyah, the Umayyad Caliph, who ruled in Damascus from A.D. 680–83, was a keen falconer and it was his family which built the hunting lodges at al Amrah and al Kharanah in the Jordanian desert. These castles, which were probably used by Yazid's successors, Hisham (A.D. 724–43) and Walid II (A.D. 743–44), again themselves falconers, are decorated by splendid paintings on the walls and ceilings showing hunting scenes and bathing parties. The strong foreign influence is immediately apparent.

A foreign accent also enters the Arabs' hawking at this time, picked up from the other cultures with which the conquerors were now in more frequent contact. It is not possible to say whether there was a strong tradition of hawking on the Mediterranean coast in those days (the Arab historian, al Mas'udi, gave Ptolemy, a successor to Alexander the Great, as the man who first trained hawks) but certainly the techniques and ideas of Persia were gaining currency. al-Ghitrif ibn Qudamah al-Ghassani, Master of the Hunt to Hisham and Walid, was given the title of *al bazyar*. This is a Persian word meaning falconer, derived from *baz* meaning a goshawk, a bird little known in Arabia and for which there is no common Arabic name.

When the fortunes of the Umayyads declined and the Abbasids rose to power, founding their capital further east at Baghdad in A.D. 762, these foreign influences were increased. In the Bodleian Library at Oxford there is a manuscript copy of an abbreviated version of the first known Arabic treatise on falconry. In the preamble the writer says:

> 'We took down this book from the library of al-Rashid (the Caliph Harun al-Rashid, (A.D. 786–809), and showed it to al-Ghitrif ibn Qudamah al-Ghassani, falconer to Hisham and al-Walid. al-Ghitrif knew it and remembered that Ma'adh ibn Muslim added to it some of the sayings of the kings and the rulers of Persia. And al-Ghitrif remembered that Michael, the son of Leon the Ruler of Byzantium, when he heard of al-Mahdi's passion for hunting and the pleasure he got from it (al Mahdi was Caliph (A.D. 775–85)), made him a present of a book on hawks which had belonged to his ancestors. And al-Mahdi ordered Adham ibn Muhriz al-Bahili, from whom he had heard some amazing Arab stories, to write a book: a compendium of the knowledge of the Persians and the Turks, of the philosophers of Byzantium and the experience of the Arabs. So with the help of God and his grace, we wrote down this book and it is all to do with goshawks, sakers, peregrines and the other hawks.'

He then goes on to list his 153 chapters which cover every possible aspect of the knowledge of hawks and the curing of their diseases. In the fourth chapter which is headed 'Knowledge of males from females', the writer seems to say, if we allow for what must be a simple copyist's mistake, that the Indians and the Persians say that while in non-predatory birds the male is the larger of the two sexes, among birds of prey the reverse is so.

Apart from striking bits of information such as these, which, if they were ever known in the desert, are now well forgotten, the main feature of the book is the importance it gives to the goshawk.* The fashion for the goshawk appears to have grown as the influence of the Arabs, as opposed to that of their language, declined in the Islamic state. In time, the accepted word for the whole business of falconry became *al-bayzarah* (from the *baz* root) and the goshawk became the favourite bird, the hawk to figure most often in medieval Muslim art. This preference existed throughout the East, through India, Korea, China and to Japan, where trained goshawks were introduced from China in A.D. 244. The trade which existed between these distant civilizations and the movement of tribes across the steppes of Central Asia in-between offer some explanation for this surprisingly united preference. The imagination stalls at the scope of the research possible: 'the goshawk as a vehicle of cultural pollination in medieval Asia' – the title seems absurd as it is probable.

The preference seems surprising to a writer in the West where it is doctrine that the longwings take precedence over the shortwings both in beauty, grace and performance. There are, of course, good reasons for using goshawks. Being equal, and more, to fair-sized sakers, they are effective hunters and can be used in enclosed country. But the goshawk and sparrowhawk both differ considerably, and not entirely favourably, from the longwings in temperament, though of all the latter the saker probably comes closest. Shortwings are extremely nervous and excitable characters and as a result can be very trying to train. They have a way of staging off-days and periods of sulking which make them maddening pupils for any but the most devoted austringer – as the shortwing's trainer is properly called.

The training of shortwings is slightly different to that of longwings. It is not customary, in the East or West, to hood shortwings. Their long necks make them awkward to hood and they do not benefit from being hooded

*An interesting insight into what was perhaps a rearguard action by conservative Arabs, is the note made by al-Jahiz (died in A.D.-869) in his *Book of Animals* (*Kitab al-Hayawan*) that Arabs are not to be seen carrying goshawks as that is the function of the *bazyar*. They will, however, carry sakers and peregrines. The only reason for this, as far as al-Jahiz can see, is that the saker is an Arab bird and the goshawk is not.

as do longwings. A peregrine tame enough to fly loose at quarry is a much wilder bird than a trained goshawk. Goshawks will not be obedient or safe unless they are thoroughly manned – that is to a point where the hood is superfluous. But when they reach this state, they are delightful companions. They combine an amicability usually absent in the more introverted longwings with an inexhaustible enthusiasm for the chase. All hawks improve with work, but with longwings there is always a risk of taking the edge off their keenness – the wise falconer ends on a good note even if it is the first. Goshawks, however, and their matchbox cousins, the sparrowhawks, are definitely the better for being hard pressed and they will give of the best till they scarcely have breath to fly. For a sparrowhawk to kill eight quail in an hour is considered no more than what is due.

Because a flight with a shortwing is unlikely to take her very far from her falconer, they are usually trained to come to the fist and not to the lure. This also, incidentally, is a discipline which helps keep them tame, as to fly in to a lure is less alarming than to do the same to a man's arm. In the field, they are also flown from the hand – they will not wait on like a longwing. They do not like long flights and so acceleration is vital and Eastern falconers sometimes help this by literally throwing them off at the quarry.

In the case of the sparrowhawk, this is simply enough done. The hawk is held in the hand like a javelin, tail and legs passing between the thumb and index finger, and thrown overarm with tremendous force. The hawks quickly get used to this launching and hold their wings in for the first few moments of flight to maximize their momentum.

The goshawk, being so much bigger, cannot be treated in this way. Instead, a plaited cord noose or 'halsband' is put round the hawk's neck and the end of the line is held in the fingers of the hand carrying the bird. By drawing this string down with the other hand, the hawk is forced into a crouching, streamlined position, like an athlete on his blocks, ready for the swing of the falconer's arm. Had she been standing erect, the force of the swing would have knocked her backwards and off balance, but crouched down in this way, she is aerodynamically ready for the off. Needless to say, in the refined world of medieval Islam, no expense was spared to make the halsband, silken and jewelled, a fitting ornament to its wearer, be its practical application never so undignified.

Apart from the goshawk and sparrowhawk, the Muslim falconers trained and experimented with other hawks which they flew at every possible kind of quarry. True to the energy and inventiveness of the age, they competed to achieve some rare or unthought of coup. The falconer to the Fatimid Caliph al-'Aziz billah (A.D. 975–96) gives instructions in his treatise on how to fly hawks at duck by moonlight. The tiny merlin, unknown to the modern bedouin falconer, was highly prized and the equally miniature hobby was also trained and flown, an art now lost in both the East and West. They knew all about the sporting possibilities of the flight at the heron which, until the open country needed for this was lost, was considered the acme of the falconer's art in Europe. The crane was a popular quarry for larger hawks and there are often mentions of training hawks to fly at gazelle.

With so much enthusiasm (the Caliph al-Mutawakkil (A.D. 841–61) allowed half a million dirhams in his budget for his masters of hounds, his falconers and the keepers of his trained cheetahs), falconry inevitably became a highly technical pastime. Though the Arabic word used for the sport in their treatises was *la'b*, meaning 'play', excellence in this form of hunting was regarded as a demonstration of man's control over other creatures.

A rich vocabulary grew up for the different types of hawk and in the Fatimid treatise mentioned above there is even an indication that the falconers used to pay attention to the hawks' comparative weights, usually thought to be an innovation of the technically minded West. As François Viré has shown, hawks were classified according to their known or supposed places of origin and their colours. Falconers today, who probably know more about hawks than most, find it very hard to be precise about the classifications of hawks of different and merging races, according to our own modern systems. Who, faced with a hawk trapped in Central Arabia, can be certain whether it is a red shahin (*Falco pelegrinoides babylonicus*) or a barbary falcon (*Falco pelegrinoides pelegrinoides*)? The different and hotly debated races proposed for the saker are entirely confusing when one sees thirty together in a room varying widely in size, shape and colour. We cannot then discount the possibility that the complicated medieval system in Arabic had its own validity based, as it was, on a vast experience unavailable to us today.

At the same time it is certain that the vogue for technical terms stimulated a wish to coin a name when none was needed and to use them when they were imperfectly understood. But the business of making correct identifications between the two systems is made even more confusing when the use of a word changes. *Saqawa*, for instance, appeared to mean the lanner in the medieval treatises but is now used by falconers in Egypt to mean a type of female saker. The one signpost in this maze, which any experienced falconer will know to treat with great suspicion, is the one which offers the names of the different types of hawk as a promise of their abilities in the field when trained.

Apart from contacts with Arab society in Spain, European knowledge of the Muslims or Saracens, as they were then known, was mainly limited to the unhappy occasions of the Crusades to retrieve the Holy Places for Christendom. For Western falconers, one great fruit of these wars was the impetus which his meeting with Muslim falconers gave to the Emperor Frederick II's project of writing his great treatise. He took back to Sicily several Arab falconers and their discussions must have had a great influence on his book. *De Arte Venandi Cum Avibus*, produced about A.D. 1247, shortly before Frederick's death, remains one of the best works on falconry. In it he puts his view that training and flying hawks is as much a chance to study the natural world as it is simply to hunt, an idea which has inspired Western sportsmen ever since and alerted them, earlier than most will admit, to the need for conservation. Frederick claimed that he was the first to introduce the hood to Europe from the East.

At the beginning of this book I said that the traditions of the medieval falconers were extinct. That should perhaps now be modified as this very brief glance at the medieval world has taken us well outside the frontiers of Arabia. The coming of the Mongols and the sacking of Baghdad in 1258 ended the caliphate and the last pretence to Arab supremacy in the Islamic state. The Mongols had their own falconers, but they were by no means bookish, and no significant Arabic works on hawking appeared after this period. We have to go to the Mughals in India and the court of Persia to pick up the story again.

But in the Arab lands, dominated by the Ottoman Turks until our own century, falconry appears to have declined except in the stronghold of the desert where the events of the outside world had never mattered much anyway. The Syrian austringers, however, with their sparrowhawks must be counted as being faithful to the spirit of that otherwise vanished world. In Iraq, Arabic speaking Iraqis flew goshawks at duck as recently as the 1940s near the Iranian border. But one of those who did so told me that he believed the game was finally up, even if the rusty knowledge was not quite extinct. In Egypt, since I wrote the body of the book, I have been lucky enough to make contact with the Tahawiyyah, a large clan of the Hinadi tribe which traces its origins back to the Beni Salim who migrated to Egypt from Arabia in the fourteenth century. They tell me that flying sakers at gazelle alongside a saluki was common practice until very recently. The older falconers among them remember training goshawks and sparrowhawks when they were young men. They use a word for the hood, *kummah*, which is not heard further east in Arabia but which was used in earlier treatises, and they are quite certain which hawk is of which sex. Furthermore, like the medieval falconers, they regularly moulted their hawks through the summer. There is no practical reason why their tradition should not go back to the descendants of the falconers to the Fatimid Caliphs in Cairo. In Tunisia, both sparrowhawks and peregrines are flown at migrating quail and in Algeria and Morocco there is still a tradition of hawking, flying mainly peregrines and lanners.

By visiting such people one can hear a faint harmony of the grander chords struck eight hundred or a thousand years earlier in the same places. Reading the memoirs of Usamah ibn Munqidh, a Syrian who hunted throughout Egypt and the Levant in the middle of the twelfth century, the distant figures who in history books are concerned with remote and strange political struggles, emerge as recognizable characters enjoying the same excitements with their hawks as we do today. Usamah, as he himself admits, was not a practising falconer but he kept men to train his hawks for him and he learned much from his father who seems to have had a more practical nature. We read of duck hawking expeditions along the river, very similar to the descriptions in Richard Blome's book, printed in 1686, of flights at hares and the problems of imping broken feathers, of the anxieties of having to sit up with a hawk all night to get her sufficiently manned for an expedition the next day, and the excitement of receiving a new bird. But above all there is the true note of the happiness of company in the hawking field which, whether outside medieval Aleppo or modern Cairo, in the wastes of Arabia or on Salisbury Plain, is a great recompense for the troubles of being a falconer.

Glossary and Vocabulary

A list of some English technical terms and other useful words with their equivalents in Arabic. There are of course local variations for some of these Arabic words, but those given below are the most commonly understood.

adj. adjective; collect. collective(ly); n. noun; plur. plural; v.i. verb intransitive; v.t. verb transitive.

Bag, hawking *mikhla*.

Bate v.i. *yikfakh* A hawk bates when she tries to fly off the fist or block, but is held by her jesses.

Beak *minsar*, *minqar*.

Beam feathers collect. *mauz*, n. *mauzah* The primaries of the wing.

Bind v.t. *yilhaq* Clutch and hold the quarry, usually in the air.

Block n. *wakr* plur. *awkar* The hawk's perch. The block is shaped like a huge wooden drawing pin with an elongated spike. The top is padded with leather or carpet. The hawk is kept tied to the block when not in the moulting box.

Bustard see Houbara.

Calling off n. *munadah*, v.t. *yunadi* The exercise in training of luring a hawk from the hand of another. The same word in Arabic is used for luring the hawk in the field.

Casting n. *rubajah* The morning after a meal in which the hawk has taken in a quantity of fur, feathers or bone with her food, she spits out a pellet, or casting, which contains these indigestible parts of the meal. The verb is 'to cast' *yirubij*. The fur or feathers given to a hawk is also called casting, but this has no Arabic equivalent.

Cope v.t. To cut the talons or beak of a hawk to remedy overgrowth or protect the hands of the falconer in the early stages of training. A practice rare among the Bedouin.

Courser, cream coloured n. *durraj* (*Cursorius cursor*).

Creance n. *khait, sabab* The long line on which the hawk is called during training.

Curlew, stone *qairuwan*, plur. *qarawin*, (*Burhinus oedicnemus*).

Decoy n. *niqil, barak* The tame hawk used as a decoy to catch a wild one.

Eagle *'aqab'* general term.

Enseam v.t. *yihqun* or *yunazzil al- shbaikah* To purge the hawk of excess fat by means of a physic (sal ammoniac or small dose of sugar) especially after the moult. See Gleam below.

Enter v.t. *yu'allim* To fly a newly trained hawk at quarry for the first time, whether or not in contrived circumstances. An entered hawk may be called 'made', in Arabic *mu'allam*.

Eyass n. A young hawk taken for training from the nest. No Arabic equivalent. Nestlings, however, are thought by the bedouin to range in size according to sex and the amount of food they can get from their parents. They are called, in descending order, *kamil, nuss, mathluth, ladhidh* and *tiba'*.

Frounce n. *jidri* Usually trichomoniasis, a disease of the hawk's mouth and throat caused by the parasite *Trichomonas gallinae*. The parasite is often associated with pigeons and hawks can contract the disease when they are fed on these. Frounce is easily recognizable as it causes cheese-like sores in the mouth.

Flags n. *khanajir* (lit: curved daggers) The secondaries of the wing.

Gazelle *ghazal*, plur. *ghuzlan* General term. *'idmi – Gazella gazella arabica. afri – Gazella dorcas.*

Glove n. *dass*, plur. *dusus*, *kaff*, plur. *kufuf* On which the hawk is carried. See *manqalah* below.

Goshawk n. *baz*, plur. *buzah* (*Astur gentilis gentilis*).

Gleam n. *shbaikah, rumuj* The grease cast by the hawk after enseaming.

Gyr *sunqur, shunqur, shungar* (*Falco rusticolus*).

Haggard n. adj. *qurnas*, *qurnas al-hawa*, plur. *qaranis* A hawk which has moulted in the wild.

Hare *arnab*, plur. *aranib* The general term. The jack is *khzaz*, the doe *'idanah* and the leveret *khirniq*.

Harrier *abu haqub* General term.

Hood n. *burqa'*, plur. *baraqi'* The leather cap which is put on the hawk's head to blindfold it and thus pacify and immobilize her.

Houbara n. *hubara*, plur. *habari* Macqueen's Bustard (*Chlamydotis undulata macqueenii*). The slimy excreta which *in extremis* a houbara squirts at a hawk, is called in Arabic *tamal*. When the houbara hangs its wings and spreads its tail in a defensive display it *yitakubir* (enlarges itself) or *yibni* (builds).

Imp v.t. *yiwassir* To mend a broken feather by putting a spine of horn, wood or metal into the shaft of the broken feather; it joins the two halves as an 'intra-medullary' pin.

Intermewed adj. *rabib* An eyass or passage hawk which has moulted.

Jess n. *sabaq*, plur. *sbuq* Made of leather or woven silk or cotton, the thongs which are attached to the hawk's legs at one end and the leash at the other.

Khan n. An eastern inn or halt.

Lanner n. (*shahin*) *wakri* (*Falco biarmicus*).

Leash n. *mursil, marbat* The leash ties the hawk's jesses to the block or the falconer's glove or *manqalah*.

Low adj. *da'if, jau'an* A hawk which has been debilitated, rather than sharpened by hunger is said to be 'low'. The opposite (high) condition, waywardness and unmanageability brought on by overfeeding, is not common among the Arabs' birds and there is no special word for it in Arabic.

Lure n. *tilwah* The bundle of feathers which the hawk is taught to associate with food. She comes to the lure when it is swung for her. To swing the lure – *yilawwih*.

Mail n. *tibn* The breast feathers of a hawk.

Majlis literally a place of sitting. An assembly or social gathering of men.

Manqalah n. Muff-like cylinder of canvas or carpet worn on the falconer's left arm instead of a glove. The manqalah is used in the Gulf.

Manned adj. *milaf, muaman* Said of the hawk which has been sufficiently carried to have become accustomed to people, animals, cars etc. and their noises. She sits quietly on the fist without bating.

Moult v.i. *yihassir, yaqarnis, yuqayyaz* Hawks moult and renew their feathers between April and October.

Mount v.i. *yihliq* To gain height above the quarry and reach a good position for a stoop.

Mute v.i. *yidhruq*, n. *khadhaq, dharaq, marsh* Droppings of a hawk are called mutes. A hawk which is about to mute very slowly lifts her tail and fluffs the feathers round the vent. Seeing a hawk which is being carried do this, the Bedouin shout '*marsh, marsh*' to alert her falconer to swing his arm and aim the bird clear of carpet or clothing.

Parasites n. collect. *qumlah* This is the general Arabic word for the external parasites, usually lice (*Mallophaga*) which are quite often seen on sick or low hawks.

Passager or Passage hawk *farkh* A hawk caught on her first migration south, i.e. in juvenile plumage.

Peregrine *shahin, shahin bahri*, plur. *shawahin* (*Falco peregrinus peregrinus*) Female – *kamil*; male – *qussiyyah, tiba'*.

Quarry n. *farisah*, plur. *farais* The game at which the hawk is flown.

Reduce v.t. *yijawwi'* To lower a hawk's condition by shorter commons.

Rouse v.i. *yunafidh* A hawk rouses when she lifts her feathers on end and shakes herself to bring them all down again in their proper places. Regular rousing is an indicator of good health and contentment.

Sail n. *raff* plur. *rufuf* A hawk's wing.

Saker *saqr*, plur. *suqur* The general term but one of the following specifics is usually preferred. Female – *kamil, hurr*; small female or large male – *mathluth*; male – *qarmush*.

Seel v.t. *yikhayyit* To sew up the eyelids of a hawk when no hood is available. A seeled hawk is *mukhayyat*.

Slip v.t. *yilfat, yihid* To release or throw off a hawk at quarry.

Snare *shabakah* Literally a net, the noose harness used with a pigeon.

Sparrowhawk *bashiq*, plur. *bawashiq* (*Accipiter nisus nisus*).

Stoop v.i. *yihawwil 'ala, yimsah, yusaqq, yadhrib* A hawk stoops when she makes a downward, perhaps almost vertical, attack at great speed on the quarry.

Swivel n. *mijawwal, mijwal, midawwar* Two small rings of metal linked by a pin which prevent the leash from tangling.

Tiercel The male of any species of hawk. The words male and female (*dhakar* and *untha*) are of little relevance as the bedouin do not identify the sexes as we do. For equivalents see under names of individual species.

Trap v.t. *yisid*.

Bibliography

Abu' Abd Allah al-Hassan ibn al-Hussain *kitab al-bayzarah* ed. Kurd Ali, Arab Academy, Damascus, 1953. cf. translation of part of this work by Francois Vire, 'Le traité de l'art de volerie', *Revue Arabica* vol. XII, Leiden, 1965.

Allen, M. J. S. & Smith, G. R. 'Some notes on hunting techniques and practices in the Arabian Peninsula' *Arabian Studies* II, C. Hurst, 1975.

'Amr ibn Bahr al Jahiz *kitab al hayawan* ed. 'Abd al-Salam Muhammad Harun, Mustafa al-Babial-Halabi, Cairo, 1938.

Cooper, J. E. *Veterinary Aspects of Captive Birds of Prey* Standfast Press, Saul, Gloucestershire, 1978.

Delme Radcliffe, Lt Col. E. *Notes on the Falconidae used in India in Falconry* Kent & Co, London, 1871.

Dickson, H. R. P. *The Arabs of the Desert* Allen & Unwin, London, 1956.

Doughty, C. M. *Travels in Arabia Deserta* Jonathan Cape, London, 1885.

Al-Hajjaj ibn Khaithamah *kitab manafi' al-tair wa 'ilaj daihim ala al–tamam wa al-kamal* Fifteenth-century copy of eighth to ninth century text, MS Marsh 148, Bodleian Library, Oxford.

Harting, J. E. *Bibliotheca Accipitraria* Quaritch, London, 1891.

Jameson, E. W. *The Hawking of Japan* University of California, Davis, California, 1962.

Mavrogordate, J. C. *A Falcon in the Field* Knightly Vernon, London, 1966.

Meinertzhagen, R. *Birds of Arabia* Oliver & Boyd, Edinburgh, 1954.

Mercier, L. *La chasse et les sports chez les Arabes* M. Rivière, Paris, 1927.

Moller, D. 'Studien zur mittelalterlichen Arabischen Falkenereiliteratur' *Quellen und Studien zur Gesichte der Jagd*, vol. X, Kurt Lindner, Berlin, 1965.

Musil, A. *Manners and customs of the Rwala bedouins* American Geographical Society, New York, 1928.

Phillott, Lt Col. D. C. *baz-nama-yi nasiri* translated from the Persian, Quaritch, London, 1908.

Schmidt-Nielson, K. *Desert Animals* Oxford University Press, 1964.

Thesiger, W. *Arabian Sands* Longman, London, 1959 and Penguin Books, 1964.

Usamah ibn Munqish *kitab al-i 'tibar* ed. H. Derenbourg, F. Lerue, Paris, 1889. cf. *Memoirs of an Arab Syrian Gentleman*, translated by P. K. Hitti, Khayats Oriental Reprints no. 7, Beirut, 1964.

Viré, F. 'Falconaria Arabica–glanures philologiques' *Revue Arabica* vol. VIII, 1961, vol. IX, 1962, Leiden.

Wood, C. A. & Fyfe, F. M. *The Art of Falconry* a translation of Frederick II's *De Arte Venandi cum Avibus*, Stanford University Press, California, 1943.

Index

Personal and place names with the prefix 'al' are indexed under the initial of the name following the prefix. Numbers in italics refer to illustrations.